# Healing the Earth - What's Love Got To Do With It?

# Healing the Earth - What's Love Got To Do With It?

Tipping Point or Turning Point? – Book 2

Sandra May Hodgkinson

**Maleny Press**
www.malenypress.com

## Healing the Earth - What's Love Got To Do With It?

Copyright © 2020 Sandra May Hodgkinson

The moral right of Sandra May Hodgkinson to be identified as the author of this work has been asserted.

This book is copyright under the Berne Convention. No reproduction without permission.

All rights reserved. Except in the case of brief quotations embodied in critical articles or reviews, no part of this book may be reproduced or transmitted in any form or by any means whatsoever, electronic or mechanical, including photocopying, recording, or by any information storage and retrieval system, without permission in writing from the publisher.

First published by **Maleny Press**, 2020

For information, please address all enquiries to the publisher via at www.malenypress.com

ISBN: 9798580412948

### Disclaimer

It is understood and acknowledged by the purchaser or reader of this publication, that the material, opinions, and experience expressed in this book are in the nature of general comment only and are not intended to be taken or used as specific personalized advice for any individual reader. The reader and purchaser will not hold the author or publisher of this book liable for any loss or consequence as a result of any action or non-action resulting from the reading of this book. The author and publisher disclaim any liability with respect of any person who seeks to rely upon the material in this book. This book is not a substitute for a reader's sound judgement and/or appropriate professional advice from suitably qualified people based on the reader's individual circumstances.

## DEDICATION

This book is dedicated to my dear mother,
Edith Irene Davies,
whose unconditional love sustained
and nourished me all my life.

My mother truly exemplified
the meaning of Living Love.

# CONTENTS

|  | Introduction | 1 |
|---|---|---|
| 1 | Perception and Violence | 6 |
| 2 | Evolution Gone Wrong? | 20 |
| 3 | The Third Matrix | 30 |
| 4 | A Hitch in Evolution's Plan | 37 |
| 5 | Patriarchy | 50 |
| 6 | Force vs Empathy | 63 |
| 7 | Mankind's Right to Dominate? | 69 |
| 8 | Kinship With All Life | 79 |
| 9 | Sacredness Within Nature | 89 |
| 10 | "Creating Your Own Reality" | 106 |
| 11 | The Perspective of the Perceiver | 123 |
| 12 | Love and the Heart | 132 |
| 13 | The Critical Role of Parenting | 142 |
| 14 | What the World Needs Now | 151 |
| 15 | The Next Step on Our Evolutionary Path | 162 |
| 16 | Our Responsibility to the Planet | 174 |
|  | Other Books By This Author | 185 |
|  | References | 187 |

# INTRODUCTION

Despite decent lives and principled and even self-sacrificing actions of many human beings, the shadow side of human nature appears at this stage in our history to be taking us on a crash course toward disaster. Daily we encounter reports of new threats to the ecosystem, unprecedented climatic events indicating accelerating climate change, as well as more violence and wars throughout the planet. As a species we present a paradoxical image; we have reached the top of the tree of life with our superior brain development but simultaneously we are destroying our environment and exacerbating climate change with our insatiable need for power. Alarmingly we are carrying this to the extent that we are threatening our own survival.

Zen poet Thich Nhat Hanh, when asked what we most need to do to save our world, answered, "What we most need to do is to hear within us the sounds of the earth crying." Why is it that so many of us are not already hearing the sounds of earth's distress? In fact why are we not all crying?

We are not all crying largely because we hold very strong belief systems and these control our perception, attitudes, world-views and reactions. Our beliefs govern whether we perceive climate change as a risk, how that might affect our lifestyle and if we personally feel safe from its influence. Outside those considerations many of us can put aside further reflection on the issue as other immediate concerns crowd in. Many of our belief systems are self/family oriented. More far-reaching worldviews can too easily be allocated a secondary place in favour of more immediate and pressing issues.

It is the fundamental worldviews we rigidly adhere to that determine the risk assessment we attribute to climate change and other threats. The beliefs we hold about the relationship between ourselves and a higher power, the environment and other sentient beings, as well as our right to dominate the earth, determine the value and the respect we ascribe to our planet and its occupants. Beliefs confine us within the limits of a specific level of consciousness. This level filters incoming information through a preconceived value system and ideology and is accepted as Truth so completely that questioning those accepted beliefs becomes challenging to one's identity.

Various thinkers have divided our consciousness into levels of awareness which define the way we perceive and react to situations. These levels hold our perceptions and reactions within bandwidths of awareness which range from irresponsible, closed-minded and bigoted to understanding, appreciating and discerning. These bands of consciousness also hold beliefs which vary from empathetic, compassionate and inclusive to narrow, competitive, and indifferent to all that is considered "other". The dominant belief system held by a race or creed influences the lives of millions of people and can determine the destiny a country as revealed countless times in the history of mankind. Ultimately then, it is our level of consciousness and the beliefs residing within the majority of the population that is the pivotal deciding factor in the fate of the planet.

The beliefs that we hold sacrosanct and which govern our perception and our responses to life are often held unchallenged from childhood into our adult lives. As a result, issues such as the current climate emergency have the power to divide people into intensely opposed factions. The same power over human consciousness is held by issues such racial bias as well different religious beliefs and political convictions. As a result it is rare for a person to examine or change his or her beliefs without a pretty powerful incentive.

This is not a particularly encouraging thought at this critical Tipping Point in our history. Having said this, it might appear strange when I say that I now have a feeling of optimism in the face of the threats to the environment and its inhabitants even when considered alongside these disturbing traits in human nature.

Why optimism? The current world situation predicting immanent disaster could provide just the catalyst required to catapult humanity toward the change we need. I believe our present predicament could better be described as a Turning point which may well serve to alert us to the consequences of our aberrant lifestyle.

This period in our evolution provides perhaps enough chaos to give us a chance of survival, but survival under different rules.

Margaret Wheatley in her book "Leadership and the New Science: Discovering Order in a chaotic World" says "anything that disturbs the system plays a crucial role in helping it self-organize into a new form of order . . . growth appears from disequilibrium, not balance,"[1]

This of course seems counter-intuitive, after all we have progressed globally through taking control of all systems in our lives. Wheatley holds the opposite view. She says we must *relinquish control*, let go of outcome and instead pay attention to the natural self-regulating process by which things evolve. A system, she says, needs a high degree of interconnectedness and cooperation, which takes us from control to a multi-dimensional web of connection. People cooperate when they are free, when able to work toward what they love, what inspires them, not because they are coerced into it. We work best when stimulated by a higher goal, one that nurtures and nourishes our spirit.[i1] This creates a self-transcendent, more mature system, one which is global in its orientation.

Our behaviour as a species to this point has been referred to by Bill Plotkin in "Soulcraft" as "pseudo-adolescent" i.e. irresponsible, self-absorbed and consumed by a sense of entitlement. This being so, the present critical defining moment in our history could offer mankind an unprecedented opportunity to leave behind our adolescence and claim our moral and ethical responsibility as a mature race of people.

The issue comes down to whether or not we will choose to utilize the most highly evolved brain on the planet, accept the obligation that position demands and transform the way we live.

We are living on the edge of possible disaster. It is a Tipping Point in our history. A Tipping Point such as this compels us to reassess the world itself and our place in it before we contribute to the deterioration of life as we know it. That is the ultimatum we have before us. The bottom line is; if we are to grow and prosper we must cast off the beliefs that jeopardise our future and undertake a journey toward a more evolved perspective.

This is no small task of course. Changing our level of consciousness requires all the commitment we can muster along with a large ration of dedication to a higher goal. It will require nothing less than a transformation of consciousness.

What is being asked of us resembles a 'rite of passage" on a global scale. For traditional peoples, progress toward maturity had to be earned. In First Nation American societies, a rite of passage or Vision Quest took the young person into the wilderness or into the underworld of their own psyche in order to find the inner strengths and qualities they required for true adulthood.

This book attempts to take you on a quest to find what we need to change and how to change it.

It will ask,

*Why is there so much violence in the world?*

*How did we lose our sense of reverence and sacredness in the natural world and replace it with self-centredness and greed?*

*Did we somehow create our own reality experience, and if so, why it is that we make such a mess of it?*

*Why is it that women have been so maligned by so many societies?*

*How can we better parent our children to create a better world?*

*What can we do to make a difference?*

This quest will start with an attempt to understand violence and the sense of entitlement that allows us to dominate the planet and all its inhabitants.

We will go on to see how various states of consciousness encourage inequality and injustice and how we can reverse this trend. We will investigate ways to change our perception and treatment of one another and the earth, beginning with parenting attitudes that can foster a new generation of young people more in tune with soul values. Finally we will explore a path from the physical and mental to the more soulful ways in which we can transform ourselves.

Our journey begins with a look at the human condition to discover why it is that we are not all crying along with the earth.

## 1. PERCEPTION AND VIOLENCE

One had only to glimpse the compounds comprising the Dacchau prison camp in Germany to realise that very dark energies resided there in the 1930's and early 1940's and were still detectable when I visited the prison in 1969. That people could be capable of such atrocities toward others beggars belief. Surely it could only be a person whose consciousness has relinquished respect for human dignity or one who has been brainwashed into the belief that a whole segment of society should be exterminated because of its "inferiority" to their race.

To walk into the confinement is to have human misery almost smother your senses. One can almost hear the screams and observe the torture inside the walls as if the phantom of its pain has lingered imprisoned just as its human residents were decades before. I tried to imagine the kinds of men who participated in the extermination of millions of Jews inside Dachau's ghastly gas chambers. Surely they couldn't have been men like my father, my husband or my brother. Surely these were fiends, brutal inhuman creatures. But no, they were husbands, fathers and brothers just like the ones I know and love.

How is this possible? Could my father have been capable, under certain conditions of the same level of brutality? Could I? And are there ways to counter these tragic circumstances in the future?  Were these young German men the product of deprived, cold, regimented childhoods?

Not necessarily.  Men, and sometimes women from every country and at every period of history have fought in battles that fostered inhumane and brutal behaviour.  Were fears for personal survival a driving factor?  Do warring factions all actually believe they have God on their side, that they are fighting for the glory of their country or that the enemy is made up of monsters that need to be wiped from the earth?

The truth is that the men of the Nazi war machine, while brainwashed by authority figures, were not any worse than any of the rest of us who have expressed fear for survival, hatred, anger or revenge against others.  We are all in this together.  We cannot judge another without judging ourselves. Very human factors contribute to negative attitudes and mind-sets.

To enter Dachau is to physically experience the negative energy field within its confines even to this day.  It is the human being trapped within a negative force field of hate, discrimination, conviction of superiority of themselves and depravity of "the other", as well as the domination of a more powerful force, the Nazi war machine, that enabled them to perpetrate the horrors that took place here.  It is as if a vortex of negative energy overtook the consciousness of these men and replaced decency and humanity with blind obedience to its power.  It takes far more than prejudice to take a family man and create a monster.

The root cause of violence is far more fundamental than environmental factors.  Violent behaviour is merely a symptom of humanity's problems.  The human condition is a complex affair; we are capable of great generosity and self-sacrifice and we are capable of terrible atrocities under certain conditions.  The problem lies with the nature of our basic consciousness and with the duality of our existence. Our monstrous nature exists alongside our nobility.

For a moment I invite you to stand outside the mesmerism of our Western belief structures, the pride in one's country and its history, the glory of its victories and its expansive colonization, the desire to "civilise" the "heathens" in conquered countries, the indisputable authority of both its religious beliefs and its political philosophy.

Perhaps we should attempt to inhabit the mind of an enlightened being for a moment and look down upon ourselves from a more lofty perspective. I invite you to see for a while with unconditioned eyes, with child-like eyes, free of our culturally imposed interpretation and perception. The history of mankind would appear vastly different from the glorious past depicted in our schools. Outside our belief system, let us look into this human condition and see if we can unravel it and understand how it has taken hold of humanity throughout our long history and what, if anything we can do about it.

Many of the kings, dictators, emperors, and warriors, previously revered as heroes and given titles such as 'the great', 'the strong' and 'the magnificent', stand revealed as perpetrators of deeds, horrific in their scale and ferocity. The leaders of the church fare no better. Many of the church's actions have been barbaric – all in the name of God.

And their mind-sets? From our newly acquired higher perspective, the aims of our heroes from the past are exposed as far less honourable ones such as tyranny, personal power, wealth and glory - at all costs.

Our heroes of the past now take on a very different appearance. One oppressor was replaced by another, each proclaiming their own merit and legitimacy, even their God-given/Divine right to rule. The patriarchal world view of the leaders determined the worth of individuals under their rule and dealt with those deemed less worthy accordingly. In order to justify their point of view, leaders alienated and ostracised those of different race, creed, colour, as well as any rival.

To raise support among the people, they used dissention to demonize and shame, converting "different" peoples into "contemptible" or "threatening" enemies, thus igniting the "righteous" vengeance of their people against the perceived wrongdoer. Their subjects, enraged by this propaganda, then ransacked, raped and pillaged the peoples they then perceived as demonic. The historical accounts have been glorified, sanitized and defended and then served up in schools as the proud past. Outside the cosmetic viewpoint, history appears more pathetic and tragic than proud and glorious.

Freed from conditioned belief systems we would have a world-view unfettered by imposed perceptions of reality. In our daily lives we could not help but discover that young humans who have had violence enculturated into them from an early age will go on to demand it for entertainment in the form of computer games, movies, books, videos and on television and other devices as a nightly dose.

Is this obsession with violence simply the result of dysfunctional minds of the creators of this "entertainment" or are the creators being dictated to by our own warped tastes? Whichever the case, is it really so mystifying that dysfunction is so prevalent in society? We have become desensitized to violence. It has become an integral part of our consciousness. Violence in its many guises is the way many relate to one another. It is a world view served up to us as normal.

Violence also has a strong magnetism for many. It satisfies our modernist need for constant stimulation. It provides a vicarious experience of power for those frustrated by powerlessness. Violence provides a thrill, drama, excitement within our humdrum 9-5 lives. It is a distraction for those who find no purpose or meaning in their existence and for those who feel society has dealt them a raw deal.

We tend to create a persona for ourselves influenced by those we admire as well as those we are most exposed to in our lives. In this way many people are drawn to building "an identity" based on screen heroes, while TV personalities are admired and idealized. Many people are also drawn to emulating methods used by business predators and ruthless politicians who hold power over others and are respected as leading members of society. Our blinkered mind-sets can too easily be led and subtly controlled by the magnates of money and media. How can values taught by parents, such as "being well behaved", showing kindness and having compassion compete? These values seem tame and unstimulating, free of the drama that is craved, puny in comparison with media representations of entertainment.

When our identity is dependent upon firmly imposed belief structures, strong emotions become aroused and prevent new beliefs getting a hearing. It is only when unprecedented events appear or when violence or extreme weather events disrupt the lives of those close to us, that a more compassionate mind-set is aroused. Strangely at times of the worst disasters, the best in human nature often appears.

Too often our adopted and enculturated goals of "getting ahead" are at odds with "soul values" such as compassion. As a result many of us suffer from spiritual emptiness, cynicism, escapism, scepticism and the absence of belief. Unfortunately these states, themselves, are the breeding ground for the negative reactions that reside within this soulless level of consciousness such as resentment, anger and violence.

The worst outcome of our unprecedented exposure to media-based violence is that it stimulates our fear. We no longer feel safe in our own streets or even our own homes. We fear that child molesters may mistreat our children on their way to school. After all, every night on television we see people being attacked and learn that villains and terrorists may strike anywhere at any time. We learn from the media that the world is corrupt, that authorities cannot be always relied on, and that ruthlessness is a proven way to survive and prosper. The fear that is aroused itself ignites our primitive fight or flight mechanism in the brain.

The result of this media barrage is that we feel increasingly unsafe and fearful. As a consequence, some come to admire the fearless, emulate the powerful and the greedy and view compassion as soft headed.

Many of us appear to suffer from a type of perceptual myopia, a distorted vision of the world. It is as if we have become prisoners, gripped and held by the warped perception of reality we are constantly presented with in the comfort of our living rooms. We repeatedly see the world perverted by a viewpoint imposed upon us by the media.

While Shakespeare and other classicists of old demonstrated that violence and power-grabbing are the tragic consequence of misunderstandings and aberrant minds, today it is more often depicted as a demonstration of bravado and manliness, or merely as the automatic and normal reaction of people when angered or frustrated. In this age we value people with prestige, status, power and ability to control. Success is measured by how well people survive in a competitive, must –win, power over, wealth-obsessed world rather than in terms of cooperative, selfless and integrative values. To achieve status it is often accepted as legitimate that the opponent be undercut or his name blackened.

You only have to turn on a television to realise that many human beings worship "celebrity". But with a new perspective, how unreal that tinsel-life reveals itself to be; how shallow, materialistic, insatiable, deluded, publicity-based and fickle. Celebrities, especially on "Reality" TV have become the modern hero archetype whose behaviour, mode of dress and even speech habits are studiously copied as a sign of young people being "with it".

Shmuley Boteach, an American Orthodox rabbi, author, TV host and public speaker, speaks of the 9/11 firemen, who, he says, bypassed all five characteristics of the classical hero. Their strength was spiritual rather than physical; they saved people they did not even know, often losing their own lives in the process; they received very little adulation and neither financial nor romantic rewards. "On the contrary it was the terrorists who fitted the bill of the classical hero. They used cunning and brute strength to destroy the lives of innocent people in return for a promise of 70 virgins apiece. Clearly, the traditional or classical hero is an ego based model, and not one of which we should be proud or one we should accept unquestioningly. Boteach adds, "I wish to be a hero of the spirit rather than a villain of the ego."[1]

It is all too easy to slip straight into the conventional world view and unconsciously block out a more expanded view that would bring perspective to our life. A big picture view of life would ask what purpose we have in being alive and together on this planet. It would require that we search our souls for answers rather than relying on the indirect culture-based viewpoints. Such questioning does not come easily to many of us; it requires us to step out of our comfort zone and look at the world from a new viewpoint.

I fear mankind has reached a high level of industrial and technological power but has failed to develop a commensurate level of consciousness sufficient to cope with this mental capability. Thinkers from the perennial philosophers to the present day have told us a developed consciousness is one in tune with soul. Many of us have lost our way because we have lost the connection with soul along with its life-enhancing values. It is as if we are instead caught up in a negative magnetic field and we cannot see anything outside of its confines. Within the limits of this negative field, our perception, fueled by fear and distrust, add power to our life-defeating attitudes. Positive energy acts within a life-enhancing energy field. Positive perception, such as the perception of the beauty in nature, is not so much inherent in the scene itself but is the result of our own sensory connection with it and the higher energetic field we bring to it.

Many human beings have been manoeuvred into believing the "Speak" of bigoted and conventional politicians, corporate magnates, media high fliers and advertisers. Brainwashed by the consensus view, we have devalued ourselves into mere information consumers. The consensus view surreptitiously holds us in its grip below the level of our awareness.

David Hawkins in "Power Vs Force" says "The really grave danger to society lies in the silent and invisible entrainment that stealthily conquers the psyche. In the process of entrainment of the public consciousness, negative values are cosmeticized by rhetoric and manipulation of symbols." The media uses skilled verbiage to create acceptance for the spurious. Standards become contorted and their worthiness and legitimacy stretched.

Pride in one's homeland too easily morphs into condescension toward those countries considered inferior; satisfaction in one's achievements becomes the arrogance or entitlement and pride in the minds of those higher on the social ladder; the need to achieve success too easily turns into the power struggles and dog-eat-dog takeovers taken for granted in big business. The energies of entrainment hold us within their bounds so that we become prisoners within that perceptual level."[2]

Ultimately then, the cause of the violence in our lives comes down to warped perception which lowers our awareness to the egoic levels rather than raising it to the more expansive soul levels. The egoic level of consciousness, being subtly brought into acceptance in our psyches excites our self-centred, narcissistic tendencies. These tendencies in turn stimulate in some of us its pathological outgrowths; power struggles, socioeconomic class stratification and exploitation of the weak, minorities and the poor along with all the 'resources' of the natural world.

It is the big picture perspective, a world view of a higher level of worth, virtue and respect, which would allow us to transcend our present perceptual blockage and warped world mesmerism. Our unsustainable way of life, built upon military aggression, vengefulness, control and the exploitation of nature ignores the needs of other species, other nations, tribes and races as well as our own future generations. This is debased perceptual blindness revealing itself behind the fog of accepted convention, a fog which has to be seen through in order for us to find our way out of it.

It seems that a circular stalemate has been created in our culture where violence arises from our failure to transcend to a higher level, and at the same time, our transcendence is blocked by our violence. Fears, anxiety and violence thrust the brain into its primitive hind brain's reactivity. The fight or flight processing of the hind brain blocks the higher heart or soul frequencies because they are incoherent with them.

Emotions such as trust and compassion, inclusiveness, cooperation and integration have higher frequencies resonating with the later-developing cerebral cortex while distrust of the world and competition are compatible with the lower frequency energy field of the defensive hind brain. This locks us into a never ending tape-loop. Instead of using the higher heart frequencies for healing and transcending, we regress into the more primitive responses of fear and vengefulness.

Zurkov says contention and violence are proportional to the distance the personality exists from the soul. "Evil . . .[is] the result of the personality being unable to find its reference point or connection to its mothership which is the Soul."[3] The personality is the tool that the soul uses to function in the physical, and negative traits conflict with these pure energies causing turbulence within the psyche which is mostly unconscious and so is never healed. The personality becomes stuck in its lost and rudderless existence, lacking real soul values, meaning and purpose and inflicting fear and violence in its wake.

The result of all this is that we have a "crisis in meaning" in our culture, one that is obvious even to those of us who are most entranced. The crisis comes down to the way we treat one another and our planet. David Hawkins says "People cherish and cling to their hates and grievances; to heal humanity it may be necessary to pry whole populations away from lifestyles of spite, attack and revenge."[4]

This smallness of spirit easily becomes meanness of spirit while its opposite, generosity of spirit allows us to arise from small-mindedness to expanded awareness capable of taking us outside ourselves and enable us to see the bigger picture. Only then can such pressing problems as violence and the global threat of climate change be resolved.

Hawkins believes the most critical realization he has come to is that a large proportion of mankind basically lacks the capacity to recognize the difference between good and evil. This takes us back to the Biblical Garden of Eden when the innocent instinctive state of the first humans, living in tune with Nature was corrupted by eating the fruit of the Tree of Knowledge of Good and Evil.

I now understand that passage to mean that the conscious rational mind has usurped a superior position in our psyches rather than using it in harmony with our own intuitive knowing. The emergence of the intellect and the clash that ensued with the already established instinctive self was from this perspective the cause of humans' corrupted condition, our fall from grace. Carl Jung lamented the loss of the magical world when humans began to use their rational left-brain. He wrote nostalgically of "that cosmic night which was psyche [soul] long before there was any ego consciousness"[5]

Certainly in the last 300 years science has dominated our knowledge in the form of mechanistic and analytical thinking. As a result we compartmentalise, analyse, dissect, and separate our world into ever smaller parts and forget its inherent connectedness. We now see life in terms of component parts rather than as systems in which every part is interconnected. This modern attitude results in alienation and demoralization, rather than relatedness and co-operation. There is a bleak sterility in the mechanistic world-view while the latter view instils a life-centred respect, inter-relatedness, and sacredness innate in all life. How different our perspective would be if we were to see ourselves in relationship with all life rather than dominating and using it as a resource for our selfish ends.

We are living in Apocalyptic times and we are part of its unfolding. The Jungian perspective of "the Apocalypse" considers it a psychic event now taking place within the collective and individual psyche. We stand at a Tipping Point, poised on a precipice and the future depends on our capacity to climb to a higher moral level sufficient to safely outweigh the lower level of consciousness that could bring about our own downfall.

Jung's version of Armageddon then, has the archetypal energetic level of self-centredness, anger, hatred and aggression battling it out within our psyches against the heart qualities of compassion, courage and dedication in a conflict of evolutionary versus devolutionary forces. Both hatred and love reside within the human being. Only time will tell which becomes the victor. Like it or not, we have already embarked upon our journey to decide the answer.

### So How Can Soul Prevail Against Lower Energy Fields?

It is a matter of commitment and awareness. If we are not to succumb to elemental negative energies within us we must begin to look honestly within ourselves, to watch the pull toward knee-jerk reactions and aggressive responses and catch them in that moment of awareness. "We must take a revolutionary step." says Jack Kornfield, one of the leading Buddhist teachers in America, "Through the profound practice of insight, through non-identification [with our thoughts] and compassion, we reach below the very synapses and cells and free ourselves from the grasp of these instinctive forces." [6] When we refuse to identify with negative states, when we no longer allow them to overtake us, we can simply watch them and choose to deny them expression.

Kornfield's Buddhist practices show how to be aware of the way destructive energies arise and function, and how to face them directly. Mostly we see merely what we believe to be there. The warped perceptions and self-justifying positions or opinions we hold, like needing to be right at all cost, must be seen for what they are, delusions.

It is only when we can transform our own inner pain and the fears lurking behind our strong opinions and "positionalities" that we can stop inflicting blame and revenge on others. It takes mindfulness and dedication to catch the moment between impulse and reaction before the anger manifests. We become reactionary when we cling to outmoded values out of fear of change and loss.

Joseph Campbell, an American writer and lecturer, best known for his work in mythology and comparative religion, once said that the first step in understanding the earthly realm is to acknowledge its monstrous nature as well as its glory. Anger and hatred exist in the mind alongside our higher nature, and mankind must humbly search his psyche and commit to making a choice between the two parts of himself warring for supremacy.

Based neither on denial nor affirmation, Buddhist psychology, as espoused by Kornfield and others, in my view provides a more comprehensive basis for our process than many other teaching. Buddhism reveals the paradox of living within the opposites in our nature. It teaches us to be in the world but not of it. It attempts to point to a way of living with peace and liberation, free from sorrow and suffering.

The "middle path" describes the middle ground where we can experience the world without getting lost in it or identifying with our emotions. Only then can we embrace the opposites, the dualities of life rather than automatically reacting to them. Far from taking the excitement out of life, as one addicted to the thrill of chasing power and victory might expect, it allows us to see anew and afresh, moment by moment. It helps us discover a path free of contentiousness, living in the Now, fully independent of the need for specific favourable conditions, [power, winning, more money, a partner in life, a better job] but instead, with stillness and equanimity despite the conditions.

Human behaviour is, in essence, a direct reflection of the level of consciousness of the person who produced it. Human beings can be pseudo-adolescent or evolved, primitive or wise, depending on their ability to attune with soul. Through a person's behaviour and choices, their level of consciousness is revealed.

When we see authentically through new eyes and clear of perceptual indoctrination, it is possible to fully engage with life in its essence. Clarity allows us to see the world of physical matter for what it really is according to wise beings throughout the ages; a learning experience. Each situation that presents itself, no matter how foreboding it appears, becomes translated through new eyes as an evolutionary opportunity, a chance for maturation into true adulthood.

An authentic adult personality sees the perfection of each situation because within each lies an opportunity for the maturation of the personality. Above all, mature human beings deeply acknowledge that, beyond the uncultured view of life, they participate in interrelated energy systems within a dynamic whole and begin perceive it with reverence, gratitude and a desire to preserve it at all cost. Within all they see around them and in every event that befalls them, they learn to look behind it all and see the hand of God.

## 2. EVOLUTION GONE WRONG?

An incident occurred during a four-month course I attended in breath therapy, or "rebirthing" as the more misleading title was at the time. On this occasion we were working in twos, one assisted another who had been distressed by some issue for most of his or her life. I was assisting an ex-nun called Jean that morning.

After a week of working on her fears, Jean had finally come to terms with a memory of her father's incest during her early childhood and youth. While facing me with our knees crossed and nearly touching, she was able to release her rage at the terror and the sense of violation he had caused her. As she looked into my face which she perceived to be her father's, her anger hit me with full force. I knew the rage was not personally directed at me but nevertheless I felt drawn into her energy. I became absorbed by the power of her distress.

After a period that seemed like hours, her rage worked itself out. She became quiet. I sat stunned still gripped by the intensity of the moment.

It was then a miraculous change began to occur. She was no longer the outraged victim. It was as if she had reclaimed the part of herself that had been taken by her father against her will. She had reached into the depths of her psyche and reclaimed her power.

I deeply experienced the transformation as I sat before her watching her face change. Once contorted with rage, it became unexpectedly peaceful and strong. She glowed as if lit from within. The whole scene became electric and everyone in the room stopped to gaze at her, magnetised by her presence. Never before had I seen such a transformation. She became radiant. Once the powerless victim, she had now an invulnerable and commanding presence. To behold her was to know without reservation the magnificence of the real nature we all possess. I will never forget the beauty of her face, or the intensity of that moment.[1]

Extraordinary experiences and events, according to Bernard Haisch in "The God Theory" support the ideas expressed in the previous chapter that individual consciousness is linked to, or a part of, the interconnected whole. He expands this to say the occurrence of extraordinary experiences demonstrates the fact that the brain determines everyday consciousness, not as a source, but as a filter.[2] Researchers agree that large regions of our brains, much of our DNA and approximately 90% of our nerve endings appear to be redundant or to have little or no biological relevance.

Researchers have now found that the nerve endings in what Haich and it is these that may provide access a wider range of frequencies and to enable more of the wholeness of reality to be brought into our awareness.[3] As we progress spiritually and rid ourselves of issues and past experiences, we begin to activate more of these neural pathways. Perhaps the yearning and discontent, depression, hopelessness and mental imbalances common in today's society, can be accounted for by our separation from this connectedness to a wider reality.

In Book 1 of this Tipping Point or Turning Point [4] series, I investigated both entropy, in which life decays and dies, and the lesser known negative entropy or syntropy which builds, re-energises and revitalises living entities. Syntropy uses a universal energy field referred to by frontier scientists as the Zero Point Field to refuel life forms. This intelligent universal energy field is designed to provide the modus operandi for life. It provides a biological blueprint whereby cells and molecules can continually die and be replaced without destroying the identity of the living entity itself

Stem cells hold the key to this blueprint. They hold the intelligent intention or memory of the interconnected bodily systems which allows them to transform into specific cells for particular organs as needed. This overarching energy field is also the mechanism behind the flashes of insight or experiences of unity but in addition gives rise to remarkable human abilities or transcendent qualities. It seems extraordinary then that in our normal states of consciousness it is unavailable to us, especially as the field appears to contain information that could advance us in many ways and even help us transition from this life.

### Has Evolution Somehow Got it Wrong?

The answer lies unfortunately with our own child development practices and enculturation itself. I have referred to Joseph Chilton Peace here as my reference because, both in my work as a counsellor with children in schools and as a student of human behaviour, I find his work provides a very valuable guide.

According to Joseph Chilton Pearce, at each of Piaget's stages of human development, the child identifies with ever larger "matrices" [literally meaning ground, model or template] from which she learns new skills appropriate to that stage. "Matrix" is Latin for womb or basis from which life is derived, the source of energy and of growth, and a safe place from which to explore the possibilities the environment offers.

The first matrix with which the child bonds is the mother, the next the earth itself, its sounds, sights, and tastes, it feel and its smells. These matrices are the plan for the optimum development of the growing child. The biological plan becomes aborted however, when content or safe place the matrix appropriate to the nature's intent is replaced by the intentions of an anxiety-driven parent and culture so that bond is incomplete.

Anything that blocks bonding with a matrix interferes with the child's developmental pathways. At birth the mother is intended to become the matrix of learning for the child to centre her attention upon. Instead, from infancy we provide in Pearce's words, "Hospitals for delivery, bottles for feeding, cribs for sleeping, playpens and strollers for isolation, day-care centres for not caring, nursery schools for not nurturing, preschools – all [of which] create abandonment and weaken the bond." [5]

A Ugandan mother, Pearce says, is so bonded with her baby she will "know" when her baby wants to defecate or urinate and she will hold the baby out from its usual resting place on her breast, and place her over a bush. Mothers in the Western world where this innate sense is lost, resort to placing nappies on the baby. The baby is seldom carried with the mother as she goes about her daily activity but instead is placed securely and in isolation in a crib. The natural bond is replaced with measures that avoid close contact and as a result, fear and anxiety are aroused within the child. Rather than strengthening the bond between mother and baby, the child/mother matrix is compromised. The baby becomes fearful and so she cries constantly, becomes restless and sometimes inconsolable.

John Bowlby, British psychologist and psychoanalyst believed that early relationships with the primary care giver plays a major role in child development and continues to influence social relationships throughout the child's life. He postulated in his attachment theory that at each stage, a safe haven [matrix] was the basis for the security of the child and separation distress affected the child even into her adult life.

At each maturity shift the child gains physical strength and abilities. At the same time a spurt of new brain growth prepares her for that new learning. This spurt is accompanied by specific shifts in the way the brain actually processes information as the brain's functioning adapts to new possibilities for growth. The whole biological system is thereby coordinated by the bonding with each new matrix. When developing to plan, the child grows in stages. She bonds and adapts to her new environment naturally and her brain function increases accordingly.

Obviously this first stage is critical to a child's development. The child at this stage cannot venture far from the mother but expands her consciousness outwards as she develops skills appropriate to the first matrix, grasping, sitting crawling and beginning to walk on tiny feet.

But the second stage has significance too as it is this step that will mean the difference between the child's later ability to feel an affinity with the environment and a respect for nature or be indifferent to it, perhaps even perceiving it simply as a resource. The manner in which we handle this stage then is either beneficial or detrimental to the altruistic mind-set.

At seven the child bonds with the earth, her new matrix, from which she confidently develops her new capacities. She gradually expands outward to find her limits with others of the family her home and the immediate community. She will be safe while she respects the boundaries within which she is cared for, protected and nurtured. After 7 she begins to develop her own personal power, but only by confidently exploring that safe place, the earth itself. [7]

Joseph Chilton Pearce[6] has run workshops for parents throughout the US for many years specifically on the crucial need for bonding in child development and the ways in which we actually block our children's natural bonding. When parents discourage a child's exploration of the world because of their own anxiety, e.g. "don't touch", "don't put that dirty thing in your mouth", a full-dimensional world view cannot develop and mature. More and more children are being exposed at a young age to social media which depicts current events in all their raw detail. Rather than being a safe place, nature and her environment becomes a danger to be overcome in order to survive.

Eric Ericson, German born American developmental psychologist and psychoanalyst, states that if each stage is handled well, the child will feel a sense of mastery which Ericson calls ego strength. If not, the child will grow up with a sense of inadequacy.

Once the earth is perceived as a threat to physical survival, the ability to interact with the unknown or deal with abstraction can be stultified. In this way an interaction with the flow of life, and cooperation with the system as a whole is sabotaged. No wonder so many of us are out of tune with the environment and all its inhabitants.

It is important to note that a child's way of processing is very different from that of the adult. Pearce points out that the child's reality needs no correcting by adult reality; it needs only the chance for proper maturation. For example, parents should not tell the whole truth to a young child about the birth of his new sibling or Santa, fairies or angels. The child's brain is not designed to cope with this level of mature functioning.

An adult who imposes her own meaning onto the child's experience prevents full developmental learning from taking place.

The adult perceives the world with mental concepts and labels riding along in tandem; perceived objects become object-plus-interpretation. The child's learning is direct, without attached concepts, and involves all-body experiencing - touching, hearing, tasting as well as seeing. Learning should at this stage be unrestricted by anxieties about her experience. While the childhoods of my generation provided relatively free exploration of our environment [after being cleared of dangerous objects] the new trend for sanitization has encouraged tender, vulnerable immune systems and unnatural anxiety.

Along with this comes the trend for earlier and earlier day care and schooling. School is introduced between 4 and 5 years, day care much earlier, sometimes, tragically, at just a few months of age. Pearce believes parents of the "magical" or natural child should delay any schooling until age eleven! In very successful educational systems such as Finland, the age of starting school is seven.

Most of today's school systems place unprepared children into an anxiety-ridden, competitive, frightening experience of school at 4-5 years of age, confine them to desks and remove them from nature and exploration. This is fundamentally at odds with their full body learning where movement, touch, sight, smell and taste are all part of their natural drive to understand their environment.

Important too is that school denies the child's natural mode of learning, play. Play is natural. During play, the child explores with her whole body and in doing so she enhances her "concrete operational" thinking. This stage is meant to be "concrete" and "operational" because the child experiences and learns by touching, smelling and tasting actual objects in her environment. In this way the child actually learns more in her first few years than at any other time in her life. Our system of education stymies this natural learning. Instead school confines the child to a seated position and is taught to receive facts more passively.

For Pearce, regular school represents a violent birth, a repetition of that earlier hospital birth trauma which he calls a symbol of brain damage, shock, intellectual crippling and depression. The child's natural way of relating to the world is through her right brain which relates with the flow of things and expresses itself through unity and bonding with the earth.

The "magical" child is at one with her experience. She does not separate her surroundings from her experience of it. School, in contrast, instils names for things and makes them into concepts rather than experiences, which further separates mind from world. Naming things splits the child's consciousness from the earth and creates a perception of the world as "out there" rather than body-world experience. The intended bonding with the earth at seven becomes a separation that will influence the way she relates with the world throughout her life.

Nature becomes something that must be "learned about" i.e. "separated from", categorized, dominated, predicted and controlled if she is to survive in a frightening world. The child's sense of awareness and safety in the world system is lost, leading to the alienation, isolation and abandonment so characteristic of mankind as a whole. This safe matrix, where interrelationships with the earth should be established, becomes instead filled with threats to be avoided or overcome. Rather than respect the earth mankind uses it as a resource.

Great educators like Montessori and Rudolph Steiner have designed school systems more in keeping with Pearce's and

Piaget's thinking. The Montessori Method includes movement, orientation to the environment, exploration, communication, purposeful activity and manipulation of the environment rather than seated instruction to cater to the child's needs at specific ages. Many alternate school systems are based on these concepts including the internationally respected Grange Primary School in Long Eaton, UK where Richard Gerver, formerly headmaster and author of "Creating Tomorrow's Schools Today", put into action his passionate belief in child-centric learning that focuses on experience and context. Schools specializing in experiential education include Sands School, Summerhill School and Sudbury Valley School, philosophically based schools like Krishnamurti and Ananda Marga schools and open classroom schools.

Steiner school education begins for the 6 to 9 year old with the opportunity to live in state of magical union with nature through pictorial forms of narrative and sympathy with nature. From the third school year the child, who now thinks in naïve-realistic concepts is introduced to the notion of the unity between the animal world and man. Nature is personally experienced with activities like planting and tending the school garden and responsibility and caring for the less fortunate.

Pearce believes a child in a typical Western society school system bonds, not with the earth but with the left analytical brain and with cultural learned values and beliefs. The brain, flooded with abstract concepts, begins to filter out all but consensus ideas and thought structure. The specific period for bonding with the earth and relatedness with it, begins to fade at about 11 and largely disappears around 14.

The child will continue to grow, but her chance of bonding with this matrix and her formation of personal power in interaction with the earth is lost. She will grow apart from nature and develop anxiety about her safety within it. Instead the child develops the need to control and manipulate it. Interestingly this separation does not occur in more "primitive" societies where cooperation with nature is essential to survival.

Had we bonded with the earth as intended, we would not pillage and rape the earth as we have done. We could not use the earth as a resource to be plundered if we were aware of our kinship with it as well as all the sentient beings who make it their home.

All biological organisms exist by interaction between organism and earth.[6] The earth, like us, is a living ecosystem after all, it breathes, protects itself by constructing radiation belts, it regulates itself and continually recreates itself. It suffers traumatic events and attempts to recover from them. The earth and ourselves are intimately linked; we evolve or die together.

Our true nature is part of this system of interrelatedness. Because we have lost our natural bonding with the earth we jeopardize our own welfare. While animals, aligned with Nature, have an innate warning system about the approach of danger, we become victims to natural disasters. As humans, we have in most cases, lost abilities that are designed to act for our welfare. Only occasional hunches and intuition remain of this potential.

Our greatest acts can be attributed, not with the brain computing alone, but with its interaction with the total life system. In contrast, disconnection from nature happens at our peril because it always creates counter-energy of destruction. Then our greatest solutions can become our greatest disasters.

## 3. THE THIRD MATRIX

I was fourteen and living with my parents and brother, John, opposite Botany Bay in Sydney. The whole family loved the beach and swam as often as we could. The beach was our playground. My dad would swim for miles parallel to the shore with very slow freestyle which I tried to copy but which threatened to drown me between strokes.

Dad and I were the last to leave one day. Dad told me to have one last swim before lunch. I tried to impress him with my progress in freestyle, which I have struggled with to this day.

Suddenly a feeling of dread flooded my nervous system. So strong was the feeling that I immediately turned and headed for shore. I had never felt so vulnerable and threatened. As I swam as fast as my fourteen year old stroke would take me I had a sense of "Shark!", not a vision, I had never actually seen a shark, but a strong sense of an approaching predator with me in its sights.

Unbeknown to either of us, a large Bull shark lurked not far away. It accelerated and lunged toward me at great speed. At this exact moment I discovered that I could touch bottom and I ran as fast as the water would allow, certainly much faster than my swimming speed.

The shark thrust itself at me at this exact time. My sudden burst of speed meant that the momentum of its strike lifted it into the air behind me as I accelerated out of its reach. It swished its tail with a circular spray and dived back into the waves, vanishing in seconds.

I was not aware of the drama behind me as I rushed to shore. Further up the beach, my father was staring at me ashen-faced. He held me tight. He told me later of the attack and said that it happened so quickly that he could do nothing. He had previously seen sharks with their dorsal fin lowered to the side to avoid detection and this could explain why my father who was watching me swim could not see the shark's approach.

That uncanny premonition probably saved me from a very serious and painful bite that day. Perhaps it even saved my life. People have lost their lives to Bull sharks. I had never had such an intuitive warning before. The beach had been my safe space, a wonderful carefree area to frolic and play, my own backyard. There had been no cause for fear there. Perhaps it is surprising then that I obeyed my premonition but I have it to thank for my safety that day.

Something in my child brain had alerted me and I had responded immediately and without question. I wonder if today as an adult I would do the same or whether my rational mind would intervene and tell me not to be so childish.

A connection with Universal energy is the third matrix.

As childhood proceeds, we diverge even further from our evolutionary plan. The next stage, one that is critical for the elevation of humanity's consciousness is a connection with the higher energies. Tragically this potential is diverted into a race toward something we call "progress".

The learning stage between 7 and 11, is followed around 15 years with exploration of the third matrix which Piaget calls the rarest of all human faculties. Pearce's terms are rather unusual, but I will attempt to explain my understanding of them.

The third matrix involves interaction with Universal energy. Universal energy is the synergetic connectedness upon which all life depends. Universal energy is a matrix designed to relate with our brain just as the mother and earth matrices were at earlier stages. This Universal energy field or Zero Point Field is, according to Pearce, intended to act as director of operations for a receptive brain. A connection with the universal realm, he says, would create for us our initial step into the final developmental stage of human potential.

Paul Bailey, author and Buddhist master, originally an aircraft electrical engineer and since involved professionally with all aspects of education, holistic health, complementary medicine, and environmental issues, has a complimentary explanation. In his book with the amusing title "Think of an Elephant", he speaks of the God Spot in our brain where, he says, a network of nerve endings "act like ion-channelling lightning rods to attract an electrical charge and connect our bodies onto a multi-dimensional holistic matrix . . . to plug us in to a super reality and the infinite whole in ways we barely understand."[1]

Generally this process is thwarted and we are abandoned in the physical and concrete. In our society, at 15 most of us bond, not with the :Universal energy field but instead with the dominant left hemisphere specializing in analytical thought.[2] Analytical thought, being a more limited frequency than the Universal field, completes the double bind. In other words the child will not bond with the larger matrix but bonds instead with left-brained thinking feeding on its own input.

Bonding with the Universal energy is designed to be of evolutionary benefit not only to mankind but to all creatures. Examples of these benefits abound in the animal world. A fox, her instincts attuned to the earth, may carry her kits on a fine afternoon to a new hastily dug burrow higher up when she senses a flood is approaching following rain upstream. Earthquake prediction in domestic animals causes them to become nervous, fowls refuse to roost, burrowing animals to leave their holes, rats and mice to leave buildings. Dogs know ahead of time when their human is about to arrive even when they are far outside hearing range.

During the Sumatran tsunami in 2004, those elephants that could, broke free of their bonds and escaped to higher ground, long before the tsunami hit shore. Bonding with this universal matrix is essential for the survival of animals. Even octopi living below Mt. Stromboli, one of Italy's most volatile volcanoes, can anticipate an eruption and escape to a deep ledge and safety well before the actual eruption.

For humanity, real power and possibility are accessible mainly in circumstances where we consciously attune with the larger field of consciousness. That field is unrestricted by the laws that govern physical reality, as is the nature of the potential energy of creation.

During WW2 the US army set up bases in Alaska. It is reported that at one stage, when an engine fault was baffling the mechanics, a technically ignorant Eskimo appeared, tinkered with it and repaired it. Like the so called "idiot savant", he was apparently following the dictates of the Universal Energy field or, as the Eskimo himself might say, of the Great Spirit.

Edgar Cayce, the clairvoyant was said to temporarily inhabit someone's body in a trance and reveal healing information to them. Phineas Quimby a New England philosopher, magnetizer, mesmerist, healer, and inventor in the nineteenth century, could evidently locate missing people in this same way and reportedly even enter a body and throw off illness.[3]

In Pearce's view psychic events such as these are not random gifts but an innate function. Some of our greatest achievements are not attributable to the computing brain alone but occur as intended, in interaction with the total life system, designed to allow one of our greatest means of expression. Capacities such as these could be used in the service of survival were they not denied in early life just as our imaginary friends were in childhood by well-meaning parents and teachers. Humanity is unaware of the benefits of expanded consciousness because our left brain takes its place.

At around 15 years of age, children become indoctrinated by parents, teachers, enculturation, education and the church with the programming considered necessary to enable them to function as responsible members of society. We learn to identify with certain principles; to be proud to belong to a country with its "glorious" past, to be born a member of a particular church, to have a PhD, to be a real man, to have social status. We are taught to identify with qualities valued by those in our own social grouping. We are enculturated to behave in certain ways acceptable to our peers, to aspire to certain goals, to treat others according to conventional standards which determine a person's worth and to understand "good" and "bad" in customary terms. It is a left brain matrix so it gives us a limited identity, confined within boundaries of accepted reality.

In this way the child identifies not with universal inclusive, all–encompassing values as intended but with a restricted, narrow and divisive sense of self. Enculturation gives our children the time-honoured groundwork upon which they can feel safe and accepted within the norms taught by their particular country or social grouping. We are required to follow the "party line", to conform to various accepted "isms", Protestantism, republicanism, nationalism, racism, or to become a cult member or a militant activist. Conforming provides security and a sense of belonging when inhabiting a world of duality. The problem is that the left brain matrix classifies people as either "us" or "them", divides those who are of your particular social class, race, colour or faith from those who are not. And it makes enemies of all those who are different.

Our failure to identify with the Universal Matrix is no small "hitch in evolution's plan" as Pearce puts it. It is the basis of our bigotry, our competitiveness, our intolerance, and our violence. It turns pride into superiority, compassion into indifference, kindness into dog-eat-dog aggression and other people, animals or the planet into useful resources for our own benefit. Identification with the Universal Matrix, on the other hand, would raise our level of consciousness from self-centred to global centred. When global centred, there is no "us" and "them", no "for" and "against", no right and wrong affiliation, no need to fight against "the other", no enemy. The whole is respected because it is our sacred home. Identification with the third matrix would elevate mankind from a vengeful, power hungry, aggressive species into one that respects and protects the planet and all its inhabitants. It would give us a global consciousness.

The third matrix would also extend human capacities in relation to a wider context. Stan Groff, an American psychiatrist, one of the founders of the field of Transpersonal psychology and a pioneering researcher into the use of non-ordinary states of consciousness for purposes of healing, says it is possible that unusual states of consciousness could access Universal energy and modify their life experience it this way. He says the mind may be able to reprogram the "cosmic motion picture projector" that limits us to the physical.[4] These abilities may not appear to be essential to life but they do suggest that attunement with Universal energy can extend our normal range of capabilities.

Unfortunately our knowledge of the benefits of this matrix for mankind is limited by our failure to access it. Notable exceptions of course, like Einstein and Mozart, prove the point rather spectacularly.

ESP, clairvoyance, precognition and telepathy are also examples of bonding with Universal energy. A Canadian biologist, Farley Mowat, relates how an Eskimo friend, Ootek, whose father was a shaman, gained an uncanny knowledge of and rapport with wolves. He could hear wolves telling one another of the arrival of caribou and alert his tribe to their whereabouts. Such capacities are vital to a tribe's survival. When living among nature, it is common that tribes without modern means of communication respond to alarm calls of birds and animals as an unconscious process. My father told stories related to him by Aboriginal elders while droving in the outback in his youth, that early Aboriginal peoples could communicate over large distances by means of "Bush telegraph", which relied neither on drums nor runners but a kind of telepathy.

There are many examples of cooperation between "primitive" societies and animals for their survival. Mankind would, in his natural state, live in a state of oneness with nature unhindered by a sense of superiority and need to dominate. In the Pacific region, dolphins herd fish into shore so that men can easily catch them in their nets, earning easy pickings in return. In the African jungle, a small bird called a Honey Guide uses a special call reserved for communicating with men, guiding them toward bee hives. The call changes as the men get closer. When the men climb the tree and harvest the honeycomb, often sustaining painful strings in the process, they share the honeycomb along with some fat grubs with the

Honey Guide. A television documentary reports that if the men don't share, the bird will lead them astray next time, perhaps to a lion's den!

In my own small realm, the King Parrots and White Cockatoos herald me quite peevishly when their feed trays are empty. My small dog alerts me not only to cars that enter our driveway long before my ears can hear their approach but can tell me by her particular calls when she needs to go outside or when she is hungry [in this case, a very aggrieved squeal] My alpacas will scream to alert me of the presence of dogs who silently predate on unsuspecting animals in pitch blackness. The calls of these animals are as distinctive to the attuned human ear as a baby girl's when she is signalling to her mother that she hurts.

And, on a global scale, if we were identified with nature itself, we would hear the discomforting sounds of the earth crying.

Nature's agenda for us after adolescence is to discover and become one with Universal energy and with the larger than human community of life. This brings our neural structures into balance with context, a wider perspective which can lead us where evolution intends. This wider perspective would allow us to leave our self-centred attitudes and embrace our connectedness with all sentient beings that co-habit our planet. We can only imagine what the future could hold if we were to attune to these higher realms.

This leaves us with the question; if we are designed to achieve a more evolved destiny, why are we floundering along using our intellect for a devolutionary pathway rather taking an alternate pathway toward expanded consciousness?

## 4. A HITCH IN EVOLUTION'S PLAN

An incident during my rebirthing practitioners' training many years ago exemplifies the experience of unity consciousness which takes us beyond the capacity of the intellect and into a higher realm.

The class was asked to dance to a piece of music none of us had heard before. We were not to follow any routine just move to the music which was hauntingly beautiful. It filled me with its magic. As if on auto-pilot, my body began to respond to the unfamiliar strains. I watched as my hands and feet moved of their own accord as if to an exotic dance. The movements resembled an Indian or Thai style of dance, neither of which I have ever studied. I remember wondering what my hands would do next and being entranced by the intricacy of their actions. My feet were taking small flat-footed steps quite unlike the strictly disciplined ballet style I had studied as a teenager. My hands turned back at the wrists and flowed into beautiful stylised movements which left me awe-struck. I was being danced!

This had nothing to do with volition. My mind simply watched as my body was moved by an energy with a wisdom of its own. A door, normally firmly closed, opened temporarily somewhere within me and revealed a new realm. It is at such times the extraordinary happens. We plug into the miraculous.

While materialist neuroscientists state that mind and consciousness are by-products of the brain's electrical and chemical processes, Mario Beauregard Ph.D., associate research professor at the Departments of Psychology and Radiology and the Neuroscience Research Centre at the University of Montreal refutes this in his book, "The Spiritual Brain". He says "the evidence supports the view that individuals who have religious, spiritual and mystical experiences do in fact contact an objectively real "force" that exists outside themselves."

Beauregard goes on to say that these experiences are commonly associated with a transcendence of the personal identity and an enhanced sense of connection to and unity with others and the world. His conclusion is that there is a trend in human evolution toward spiritualization of consciousness. He proposes a new scientific frame of reference to accelerate our understanding of this process and significantly contribute to the emergence of a unity type of consciousness. [1]

Despite the high incidence of religious and mystical experiences reported in the US, they are still considered the exception rather than normal human levels of consciousness. More information from brain biology may explain a hitch in the evolutionary plan that prevented our connection with universal consciousness from developing into the norm, except for rare and beautiful moments such as those during my rebirthing training.

I quickly read up on the workings of the brain. The results are fascinating.

Our brain is actually a threefold neurological system; the oldest evolutionary system or reptilian brain [brain stem], old mammalian brain or limbic system and new mammalian or neocortex. Each system takes us to more complex operations and operates on different frequencies. The old brain is responsible for our instinctual flight or fight survival mechanisms, the old mammalian is our limbic or emotional system while the neocortex is responsible for our forward thinking and analysing.

The new brain is designed to reign in conjunction with the rest as a whole interrelated unit. The brain is an incredibly complex organ of interrelated structure, function and connectivity. The more we consciously develop its functioning, the greater the range of frequencies it is able to access and the wider its experience of reality.

These three systems, when integrated, should offer us open-ended potential, an ability to overcome the constraints and limitations of our lower brain. When that integration fails, our mind becomes dysfunctional, our behaviour anxious, striving and warlike and we act against our own wellbeing. Our new brain, instead of aiding our evolution, ultimately becomes devolutionary.

It is the two later-developing brain systems that are capable of experiencing the higher frequencies of the subtle and causal fields, so they are a higher evolutionary achievement than our older brain. Our higher capacities are easily overridden however, by the more stable and established old brain habituated into our system over the whole history of human life on the planet.

Our evolution's aim, Pearce says, is to take us beyond separation and toward inter-relationship with Universal energy thus enabling our brain to be able to access what he calls "unity thinking". Unity thinking involves creation with Universal energy [the Zero Point Field] and thus involves insight intelligence or inspiration from higher sources as we have found from other research.

The really exciting bit is that when we use unity thinking our perception is changed; we begin to perceive every event in our lives differently. Our lives become part of the creative unfoldment of life itself in conjunction with Universal design, a purpose in which we ourselves are deeply involved.

We might fear that our hindbrains are in control and that this leaves us with little hope for the future. But Bernard Haisch suggests there is a self-organising character to living things so that despite free will and chance, a higher order purpose is behind all of life, guiding it toward higher evolution. The imperfection we observe around us indicates only that this is a "work in progress" i.e. evolving and that humans, are part of this process.

The Jesuit Teilhard de Chardin [May 1, 1881 – April 10, 1955], a French philosopher who trained as a paleontologist and geologist and incidentally took part in the discovery of Peking Man and Piltdown Man, proposed that evolution and Intelligent design or Divine purpose are not incompatible but that our consciousness is evolving toward final unity with the Divine. Many others now believe that Divine purpose and evolution are not contradictory as often believed but are in fact complementary. Evolution exists guided by Divine purpose and Divine timing -despite the fact that we might wish it move somewhat faster.

All these authors agree that mankind will mature to full adulthood after adolescence only when our thinking process becomes open to the awesome power of Universal Consciousness. In these troubled times, however, it appears that we are not developing the higher levels of nonphysical thought we must have for the true intellectual maturity that was intended.

This failure to mature as a species would indicate either that evolution's plan has been aborted by our own actions or that our higher systems have not been developed by enough people over a long enough period of time to become stable, or as Pearce explains more scientifically, "The capacity for entering this higher evolutionary realm of potential isn't enough established to be statistically available." [2]

The trouble is the brain resembles a data processor in the physical but as we have seen, intelligence and consciousness reside elsewhere.  When the brain filters out virtually everything other than immediate consensual reality, we mistakenly believe that our restricted consciousness, governed by duality, represents the complete picture.  It is only when occasionally a "crack forms in the cosmic egg" and we break through the mental filter by means of meditation, beautiful music, an awesome scene, or even trauma, that a wider reality breaks through.

With the aid of such moments as well as life experience, a percentage of people do rise gradually to higher levels of moral development and wisdom and evolve as spiritual beings.  Many others do not appreciate such moments for what they really are and shrug them off.  Nevertheless such occasions do reveal that there is a greater nature pressing to be born within us and many of these events, including near death experiences, appear to be increasing in number, or at least are reported more often and are now taken more seriously.

Our forebrain has undergone an enormous expansion in size in the comparatively "recent" past and this has occurred apparently in the absence of any survival need.  In less than 4 million years, a relatively short time in evolutionary terms, the hominid brain grew to three times the size it had achieved in 60 million years of primate evolution.  Nature does not expand in a superfluous manner, natural processes are designed for evolutionary purposes.

In the time of Homo erectus, the development of language was attributed to a sudden increase such as this. Some researcher's associate sudden development of language occurred with our transfer from primitive pursuits such as hunting to a more settled lifestyle. Both increased brain size and the development of language gradually encouraged a search for higher meaning.

We are also questioning our belief systems. Spirituality has evolved similarly over time from early man's superstitious practices, to religious ritual, to become more and more, direct spiritual experience. The latter development in brain size has, it seems, coincided with the development of a circuit of cells in the brain called the 'God spot". Paul Bailey, who coined this phrase, calls this a transcendent faculty which is evolving in life's gene pool spearheaded by humanity. [3]

Bailey agrees this growth spurt is "designed" to take us beyond created reality and that the stimulus for it came from the Field of Infinite Energy itself. He likens this process to the miraculous moment of conception when a dimensional leap of similar proportion occurs as cells from physical reality take on a life force of their own from beyond, which he calls a "field of infinite energy in perfect balance", or Universal energy.

In the past, traditional science and the perennial wisdom philosophies have differed markedly in their approach to life's purpose. While some philosophers such as Ken Wilber and Stephen Phillips argue for the fundamental similarity between eastern empiricism and contemporary science and recognize a wide range of extraordinary experiences normally beyond our mental capacity, scientific specialization separates these approaches and makes it difficult to develop a more inclusive and broad synergy. In the area of brain research, however, just such a synergy with another avenue of study serves to give us a more complete picture.

The Institute of Heartmath in Boulder Creek, California, which we visited while living in California, employs many scientific minds to study and measure the electromagnetic frequencies within the human body. Their research indicates that higher frequencies must have a corresponding highly developed system within the physical to allow them to actualize or express themselves. In other words, the body would require a mechanism that could expand its normal functioning to become coherent with the higher frequencies.

Scientists at Heartmath have found just such a mechanism. They found that when the brain and heart frequencies entrain or work together as a synchronous whole, they form a resonant, coherent wave pattern which is critical to actualizing or embodying these higher frequencies and therefore also to the full development of the human being.

The scientists have also found nested energy systems ranging from the atom to universal frequencies which are centred on the heart. Put in in layman's terms, far from being simply a blood-pumping device, the heart appears to the focus for electromagnetic energy within the body! Heartmath scientists measured the heart's frequency and found it was actually stronger than that of the brain. The heart has in fact, the only frequency mode sufficient for transmitting higher energies into our human realm. The heart, then, is the hub of Creator-creation expression!

The heart's intelligence was found to be neither linear nor verbal such as the brain's, but holistic, unified and connected with the whole creative process. As such, this intelligence is responsible for our wellbeing and for reaching our potential. This is the intelligence that powers the God spot as part of our next evolutionary development.

It is only when we quieten our brain's negative chatter that we can we access the more subtle heart energy. Fear, on the other hand has a vibration that blocks higher heart frequencies because it is incoherent with them. Fear builds up in a field effect which locks in our receptive brain. As a result we are unable to allow full interaction with the higher frequencies.

At the root of our ever-present crisis is this failure to employ our heart-brain entrainment in the way it was intended. This finding by frontier scientists echoes the teachings of all the great saints and spiritual gurus throughout history who say that the heart is the key to the transcendence that has been in the process of unfolding for millennia. A heart-brain bonding acts for our wellbeing because it instigates "unconflicted" behaviour, or behaviour free of polarity; neither reactive nor defensive. It is this connection that provides Joseph Chilton Pearce's' "crack in the cosmic egg". It is the secret that facilitates our initiation into a higher level of consciousness.

The new evolutionary thrust is designed to take mankind beyond being successful merely at Darwinian survival traits of competition, conflict, and superiority based on physical strength, and toward an inter-connectedness based on an individual's access by the heart to the creative forces themselves. How remarkable is that!

This new evolutionary breakthrough will be brought about only when sufficient numbers of humans reach and maintain this level of higher consciousness.

### The Benefits of Higher Consciousness

The higher, inter-dimensional level that unity consciousness provides allows us to open to vast mystical states including higher spiritual realms Eastern mystics call "enlightenment". It brings subtle awareness, sensitivity and more refined thought processes, perception, and memory. It also involves access to life-giving inspirations and intuitions beyond our normally capacity. These extraordinary kinds of knowing, centred on the heart, would enable us to inter-relate with the world with greater compassion, creativity, insight and Intelligence, all qualities sadly in short supply in our troubled world.

The heart's frequencies change both our perception and experience of reality. The right brain/heart system is associated with maintenance of the whole, with integration and cooperation, requiring the parts to act in harmony and unity rather than in separation and conflict. This energy is associated with the Feminine or Yin principle. Thomas Aquinas believed that there is an intelligence related to the acquisition of virtue that is independent of abstract logic or science. He referred to this nonverbal intelligence as a kind of instinctive "knowing" which he believed to be the basis of moral experience.

As Jack Kornfield says, when this inner awareness arises there also comes rapture, happiness, expansion and the ability to enter into the profound states of silence which he calls oceanic rapture and expansiveness.[4] Critically, it is only with these more balanced integrated heart and limbic frequencies that we hold the key to transcending the violence and destructiveness of the unbalanced left-brain intellect or Yang principle.

The famous Swiss psychiatrist Carl Jung explained these experiences of unity consciousness in terms of 'meaningful coincidences' or events that bring meaning to certain events in our lives in a manner that suggest they are governed by something beyond mere chance. Such events, he says, indicate that there is some guiding quality of the world that brings the individual toward further personal growth and their deepest purpose. "We are turned, at least for a moment, beyond our usual thought and behaviour toward a greater identity where . . . we rise beyond ordinary biography into a greater calling, a larger self that appears to be joined with the creative force that animates the world at large." [5]

Already we see widespread droughts, unprecedented fires in the Amazon, Indonesia and Australia, tornadoes of a ferocity previously unseen, huge floods and loss of life, both human and animal, all due to climate change. It is only with attunement to Universal energy that mankind can collectively and globally revolutionize our self-centred approach to life and take our suffering planet from the brink of devastation.

With beings capable of such higher perspectives, evolution must advance as it was intended. This becomes possible because, deeply ingrained within this ego-transcendent nature is the development of the need within the individual to help further the world's advance

Examples of advancement such as this are the civil rights reforms and environmental and women's movements, the ending of the cold war without bloodshed, the demolition of the Berlin Wall and the elimination of Apartheid in South Africa. The type of knowing that brings about such radical changes involves a high level of emotional intelligence that allows us to overcome petty prejudice and employ a more tolerant and all-embracing world view, a crucial element in our ability to act virtuously.

Individual personal growth is required in order to enable us to become aware of and overcome their habitual ego-based patterns and as more people gain the higher perspective that characterizes the transcendent identity, a tipping point can be reached. As self-knowledge grows so does discernment and wisdom.[6]

The cultivation of our higher attributes is not an easy task; after all it means attempting to overcome our natural tendency toward self-cherishing, self-aggrandisement, and self-defence. The payoff however, is that it can create a higher perspective and detachment through which we can more wisely embrace new challenges and opportunities and come to terms with who we really are and what we are meant to do during our incarnation on this planet.

It is this Intelligence which is the foundation of the critical matrix shift that Nature designed to occur at around fifteen years of age – the shift that rarely occurs. Without it, the adolescent instead regresses and too easily becomes locked in a tape-loop of drama, of resentment, entitlement and contentiousness which can be self-perpetuating.

In the adult this trend can go on to become self-defensiveness placing the needs and desires of the self above the needs of others, power games, domination and occasionally violence enacted in varying degrees and over most levels of society. Because violence and environmental destruction is self-based, compassion and care for the environment become secondary.

Pearce's comments on this failure to reach the final matrix are telling. "Were our intellect serving an intelligence which moves only for well-being, not one facet of our current personal-social-ecological crisis could exist. The intelligence of the heart never solves problems, it just dissolves the situation in which problems exist and gives a new situation". [7] The pueblo chief, Ochwiay Biano once told Carl Jung that white men were 'always upset, their faces lined with wrinkles. . . . a sign of eternal restlessness." [8] The chief said white men were crazy since they thought with their heads, whereas it was well known that only crazy people did that. Indians thought with their hearts.

The great Indian philosopher, Swami Muktananda says the heart is the true seat of the mind. He speaks of the "primordial principle in the cave of the heart". It is within the heart's implicate order frequencies that is found the fourth and final body of yogic psychology, the realm of insight–intelligence, the witness state, the final part of us to be developed.

The purpose of this heart/mind bond is to facilitate our evolution toward unity consciousness. Bernard Haisch in "The God Theory" weighs up our chances of achieving such an elevated state when he says, "Teilhard [de Chardin] leaves us with a tantalizing implication of development on a cosmic scale, an evolutionary hope that the universe will achieve some kind of ultimate eschatological perfection rather than descend into a final state of maximum entropy." [9]

Lynne McTaggard in her book, "The Field" concludes from her extensive study of the work of frontier scientists, that an attunement with the Higher Mind "would help us take a final evolutionary step in our own history by at last understanding ourselves in all of our potential."[10] Science and spirituality are shown more and more to be profoundly intertwined. The implications of such a synergy bring us hope when our own history seems to indicate that our future is in the balance. We do indeed stand at the Tipping Point.

The question remains; are we prepared to extricate ourselves from tape-looped thinking and attune instead to unity thinking? If so, our present crises could be averted, including violence toward one another, greed-based destruction of our habitat and its inhabitants, the use of force to make life conform to our wishes and the use of power to assuage our personal fears.

In order to achieve this, much of our belief system about power and force being the natural scheme of things in a patriarchal world may need to be revised.

## 5. PATRIARCHY

Driven in my late teens by an idealistic desire to live the heroic life, I made a choice to "save myself" until my hero appeared, a dashing and amazing being who would ride with me astride behind him, [naturally a proud white stallion was involved here] into an adventurous and praise-worthy life. This naivety and Pollyanna-like quality I can only explain now in terms of long years of cloistered ballet training and a strict family background where the complete absence of television and telephone created isolation from my peers.

Extraordinarily my hero did appear in my early 20s. It was during my first year of work, the year that would provide just sufficient cash to fulfil my dream to make a grand tour of Europe followed by residence in Canada where I would teach. Then a British Royal naval destroyer pulled in to Sydney Harbour and off walked a handsome, older, widely travelled adventurer who added the finishing touches of having a dashing uniform of a lieutenant commander, complete with decorations for bravery by the Queen, and, just to give the final touch, was a swashbuckling sword-fighting champion. I stood no chance at all, and wouldn't have taken one if offered. My infatuation was immediate and, more incredibly, seemed to be returned.

After a whirlwind couple of weeks holiday with him [and, of course, my parents who were charmed by his gentleman's manners], he returned to his base in Singapore while I embarked on a ship's voyage in the opposite direction via Panama to England. Due to unforeseen circumstances, however, his destroyer arrived in England unexpectedly weeks before ours. During this time he was victim of a serious collision which gave him head injuries requiring admittance to the Royal Naval Hospital as well as lasting complications I was not to fully comprehend until later.

Naval officers have qualities which make them capable of controlling men and military ships in emergencies. It's what they do. Their decisions are not argued over or debated. They had Her Majesty's authority to get the job done in all circumstances. At the same time, the British naval officer had the reputation of being an exemplary gentleman – at least, while mentally undamaged.

From the time we met I knew that this man was to be the hero into whose arms I would happily surrender my girlhood and launch myself into true womanhood. When I met him again in England however, I sensed something wrong, something I could not quite identify. It was as if someone else now lived inside the man I had previously hero-worshipped. I decided not to rush into any relationship until I understood my hesitance.

As I was newly arrived in a strange land, he took me to a hotel [where I had insisted on separate rooms] and went off to get me a hot drink to settle me for the night while I changed and got into my bed. He returned, dropped the drinks and rammed me against the bed, held my shoulders and roughly took my virginity and along with it my trust, my innocence and my childhood.

The next day I found I could not speak. He didn't notice because he wasn't speaking to me. He had concluded wrongly that I was not a virgin and apparently felt betrayed. Although never discussed, this seemed of over-riding importance to him. It was unusual for girls of twenty-four to be virgins even at that time but apparently it was part of his criteria. The irony is that he never knew that I was.

My own sense of betrayal went much deeper. It resonated with a deep sense of the subjugation of women that lies within every woman's psyche and handed down over eons through rape and brutal persecution. For years after that night I dreamt that a murderer was trying to break down my door or that I was being held for a crime I did not commit and about to be punished in an unthinkably cruel manner.

### "The Unbonded Male Rapes." [1]

Most young naval officers of that time were drawn from the upper classes but, largely because of their "privilege of birth", they were denied the natural bonding processes. They were sent [for their own good of course] at an early age to reputable boarding schools to give them an advantage over their "less fortunate" siblings. Here they were treated with discipline, rather than the warmth they would have received at home.

We have learned that when bonding is aborted at any stage, successful maturation to higher stages is retarded. Many researchers believe the male to be functionally more affected by the lack of bonding than the female and more cut off from the life force as a result. Little boys who are neglected, abandoned or whose mothers die, suffer more lasting scars than little girls. Baby boys also appear to be less resilient than girls. It is a fact that 80% more male babies are born blind, deaf, malformed, die of crib death, are autistic, schizophrenic, hyperactive and dysfunctional than girl babies. Even our healthy baby boys, however, tend to be coddled less than girls, perhaps in the false assumption that they are stronger.

As we have discovered, each matrix is designed to be a home, a safe place where the child has a source of personal power and potential to call upon for his or her stable development. According to Pearce, the unbonded male, lacking the connectedness with his first two matrices, his mother and the earth, can in later life subconsciously attempt to gain from these matrices what he lacked in childhood, nurturance. "From the beginning, Pearce says, "the female being of the base-line genetic structuring of life is able to flow with, bide her time and survive, the male is anxious, tries to fight against, dominate, pit himself against the odds. . ."

To dominate the matrix, Pearce believes, becomes his passion. Perhaps this explains the male need to control and do battle with the earth rather than interacting with it and responding to the flow of events. It may also account for the fact that some men rape women.

Given these facts, it is not surprising that the vast majority of domestic violence and sexual assault is perpetrated by men against women and children. We live in a society where, even today, patriarchy determines the worth of human beings. In all areas we still place more men in positions of power over women [family, work, religion.]. Society at the same time presents violence in the media as a preferred means of gaining control. Our culture values those with power. Most of us have been taught that patriarchy is a natural, almost God-given condition.

Notwithstanding the fact that there are far more male heads of corporations, leaders of countries, geniuses of all fields, scientists and inventors, there are also more men at "the complete catastrophe end" of the spectrum. More males die of suicide or homicide. They compare poorly with females after divorce. There also are many more males who are alcoholic, homeless, convicted of serious crime and there are a very much larger percentage of men in prison. Beneath the masculine posturing and bravado, perhaps at some level males realise their vulnerabilities and become more subject to anxiety than women and so live in a state of fear, competition and struggle. These characteristics are after all, the foundation upon which patriarchy is able to dominate.

Despite Jesus' apparent equal treatment of women, Christianity reinstated patriarchy after his death and established the authority of the priesthood. Women became demonized as seducers of Adam and therefore responsible for "original sin". The patriarchal church is also responsible for the death of at least eleven hundred women, some reports suggest many times more, accused of being "witches", alongside a handful of men. It also sanctioned the massacre of men, women and children, all "infidels" in a series of bloody crusades. It was responsible for hunting down and burning many of its own, such as the Knights Templar, who were accused of transgressing the accepted beliefs sanctioned by Rome. Patriarchal societies do not tolerate dissention.

Jeremy Griffith in "A Species in Denial" has an interesting theory about how patriarchy came about. He speaks of an instinctive memory in human beings of a matriarchal, cooperative, alienation-free ancestry, where individuals behaved in a selfless, cooperative manner toward one another. He likens these qualities to soul where we cherish ideals of mankind behaving selflessly and cooperatively; an ideal that has long ago been lost. [2]

Griffith blames the full development of the conscious mind and its search for knowledge for the corruption of that innocent instinctive state. When men embarked on their search for knowledge, the genes followed, selecting for individual left-brained thinking with its innate egoic tendencies rather than the hereditary or instinctive right-brained more inclusive, matriarchal approach to life. Divinity was separated from knowledge and reason from matter. Thus, in the drive toward the knowledge of good and of evil, instinct, innocence, nature, the senses and spirit became synonymous with ignorance and paganism.

Eastern religions such as Buddhism and modern psychologists following Jung lament this unbalanced development, speaking of the world of psyche or soul that was lost with the development of ego-consciousness.[3] If the unbalanced conscious mind rather than original sin is, in fact, the culprit enticing man into pursuit of knowledge, perhaps Eve can be exonerated after all.

Along with ego-consciousness came competition and survival of the fittest, which in latter times has developed into our power-fame-fortune-glory mania, ending forever the integrative, cooperative living and altruistic values of the former matriarchal society. In spite of this, Griffith attributes no blame to men for taking the path toward patriarchy. "Since ignorance was a threat and since men were the group protectors, it was men who had to take on the task of championing the intellect and defeating the ignorance of our soul. With this change in priorities, humanity changed from being a matriarchal society to a patriarchal one .."[4]

The priorities of women, on the other hand, have always been with child rearing, child nurturing and safety rather than competition and so, according to Griffith, they had little sympathy with the battling, aggressive drives of men. In this way men placed themselves in a position of power over the group and domination over the women. With the subversion of matriarchy and innocence however, women were forced to suffer the destruction of their previously more soulful lives.

Griffith again, "Until men could explain to women why they have had to be so egocentric, competitive and aggressive there could be no fundamental change to the situation where men found they had no choice other than to oppress women ... While men have yearned for freedom from their oppressor, ignorance, women have similarly yearned for freedom from their oppressors, men."[5] As Griffith explains, the tragedy was that, as the battle of the sexes developed, neither could be labelled bad or good, both were being heroic in their own way and under the particular pressures they were experiencing.

This chapter is not intended as an indictment of men or even of patriarchy itself; judgment is never a path leading to a solution. Let us instead attempt to understand patriarchy and to find some way of creating balance and justice out of that understanding.

Patriarchy apparently began in the second millennium BC, in the Bronze Age after a period of 500,000 years, referred to in many traditions as a time of Eden, in which the land provided for every need and their social and religious lives were rich and complex. This early "matriarchal" society was so named because it offered those social relations most suited to the life of human groups and communities that were closely connected with nature and assured the nurturing of the young as well as the groups' or communities' survival.

These egalitarian societies of ancient Egypt and Mesopotamia were not so much led by females but lived in tune with Nature in a more mythic sense rather than dominating it. Hunter gatherers lived in harmony with nature; they hunted deer for food, used bone for tools, made clothing and shelters out of hides. They moved with the seasons. They were part of nature rather owning it or controlling it.

Many writers believe that early societies lost their matriarchal character when climate changes forced huge migrations of Aryan nomadic herding people which caused dangerous conditions for established peace-loving families.

My sense is that the change from the matriarchal society began much earlier around 4,000BC, along with a giant social change that brought nomadic tribes with their seeds and their animals domesticated from wild stock, displacing the hunter gatherers of the Mesolithic age and establishing the Neolithic, or new social structure. These newcomers brought in the concept of ownership of land which changed the lives of the matriarchal peoples and started mankind on the race toward modern society.

Archaeology in the UK where tribes from the Continent brought with them the Neolithic era, reveal flintheads of a kind used for weapons as well as deeply wounded skulls, indicating that ownership meant conquering native people and taking over their hunting ground. These conquerors controlled and enclosed farmland and built homes surrounded by stone structures and deep ditches. The population rose dramatically from around 2,000 to 100,000 over this period and competition for the best land became fierce.

Thus, as history unfolds, the conquering male took over the property of neighbouring tribes, confiscated cattle for food, prisoners for slaves and women for extra wives, thus further raising their own wealth, power and status. It was the technological advances of the Bronze Age, coupled with climate changes that intensified warfare and the resulting unstable conditions that meant it was advantageous for women to have the protection of men for themselves and their children.

Despite the fact that patriarchy provided a solution for both at the time, protection for women was bought at the cost of their freedom. Anthropologists like Levi-Strauss give another point of view. By marrying their women into another tribe, some clans might actually have become pacified. In either case, women became the pawns of men.

In their wish for a male to inherit his status, men overthrew the old female order of inheritance and replaced it with the patriarchal system. Jewish, Hindu, Chinese and Christian and Greek ideologies all defined women as subordinate and established rules to enforced this. Instances of injustice toward women abound. Women, for example, could not hold any public office but were imprisoned in the home. Ancient codes of law punished female adultery severely, while not concerning themselves with the philandering male. Women's property was likely to be seen as dowry to attract a husband, whose property it became and generally, women could not appear as independent persons in court.

In some countries and over many centuries, including parts of Russia, it has been considered proper for a man to beat his wife regularly as a precautionary measure. English law enshrined the right of a man to beat his wife provided the rod was no thicker than his thumb . . . a fact that led to the saying, "the rule of thumb". While this appears to our modern Western view to be indefensible, these men were merely living up to values expected by their culture.

Our history, recorded as it is by the conquering patriarchy, is a record of domination and war. The great and often quietly heroic, certainly stoic achievements of women against all odds, go largely unrecorded and unacknowledged, like the bonding that apparently occurred in Changi and other occupation camps. Similarly women who form a sisterhood involving intimate exchanges which allow emotional release, have for centuries maintained a resilience which has withstood the oppression of the men who hold superior positions of power over them, mentally, emotionally and physically.

While many cultures believe that patriarchy is the natural order, the subjugation of women, the bearers of life-loving energy, has the effect of diminishing and threatening all life. As Barbara Hope says, "Life for women, life for the earth, the very survival of the planet is found only outside the patriarchy." [6] The suppression of intuitive wisdom-traditions and the colonial destruction of indigenous cultures, oriented as they were toward inter-relationship, caring and nature-worship, was often violent and barbaric. This destruction distorted natural creative expression when cooperative co-habiting was transformed into fear for survival. Fear, as we know, inevitably leads to violence, conflict and injustice. "Love and devotion are vibrational frequencies that maintain reality [but] Love can only be given in freedom." [7]

Today, in vast areas of the Earth, women are still in bondage, treated like slaves and denied their basic rights. In India some are still burned alive over dowry disputes. In China female babies are often sacrificed because of a wish for a male heir in their one-child-only society. The scourge of discrimination continues around the Islamic world, a practice where women are murdered for the smallest affronts to rigid social or dress codes. An unproved charge of sexual infidelity or even a husband's paranoid dream may result in a woman being burned or hacked to death. Her murderer receives the approval of family and community for his action.

The rape of a young girl, sometimes even by a member of her own family, caused the family to believe she has brought dishonour to the male members of the family who then feel justified in killing the girl, in what is known as "honour killings". Along with genital mutilation, forced marriage, the denial of education for girls, and unequal "hadd" punishments, unspeakable injustices continue to be perpetrated against women today.

Patriarchy is about power, power over others. Patriarchy is also synonymous with war. Historically, patriarchy began with war, developing as we have seen when armed men on horseback raided towns, looted and conquered their people, and by enslaving them, made themselves more powerful. In this process the whole disaster of social discord arose – class systems, property theft and ownership, slavery, hostility between the sexes. Individual as well as collective violence and injustice, from the king down through society, is always the result of hostility and domination. And so war creates war in a continuous self-generative process.

According to R. Eisler 1993; M. Mies 2003; and C. v. Werlhof 2003, wars are nothing short of organized killing presided over by men. "They have absorbed, in the most complete way, the violent character of their own ethos." They design missiles and technologies which reinforce their dominance, weapons ostensibly to create peace but which are capable of destruction on an unprecedented scale. "These are the men honoured as heroes with steel minds, resolute wills, insatiable drives for excellence, capable of planning demonic acts in a detached non-emotional way".[8] Patriarchy in its extreme produces what Werlhof calls 'dead men, the hollow men", capable of extraordinary excesses of violence.

Claudia von Werlhof goes on to say that after many periods of human history in which matriarchy prospered, patriarchy is a recent evolutionary error and a very violent and dangerous one at that. Our social order as a global patriarchy has evolved over a period of 5 to 7000 years but, she believes has reached its final stages. Certainly this globalization and domination of the earth by machine is unsustainable and the patriarchal combat with competitors and destruction of nature holds the possibility of destroying life as we know it. Human beings took an expedient step, thousands of years ago, and over millennia have turned it into a potential holocaust.

## Patriarchy vs Evolutionary Design - Is There Any Hope?

If patriarchy is yet another error created by mankind in opposition to our evolutionary design, where does our hope lie? The answer may lie in brain physiology as mentioned in the previous chapter. While patriarchy predominately lies in an imbalance within the left brain, the right brain could offer a solution to this imbalance. While the Masculine left, analytical brain causes separation and alienation, and can act without Insight-intelligence, the right brain's pre-frontal lobes, usually associated with the Feminine or Yin, [and not necessarily with females] give direct access to the subtle, more balanced integrated heart energy. This is the key to transcending the violence and destructiveness of the unbalanced Yang intellect.

The right brain is associated with maintenance of the whole, with integration and cooperation, with harmony, and with unity rather than separation and competition. Only the connection with the Feminine can reunite us with the wisdom that has been lost. In destroying the feminine, intellectually based and unbalanced male energy has taken the Earth to the brink of devastation; in reuniting with the feminine, Yin energy, humanity has a key to living in peace, harmony respect for one another and the earth.

I am not suggesting a return to a matriarchal society but more to a balanced Masculine/Feminine energy. The Feminine Yin qualities of love and compassion are powerful energies, energies capable of dissipating the power-over, dominating and forceful Masculine Yang energy. Pearce gives the example of a young woman abducted by a group of violent young men who pushed her into the car and told her they were going to rape and kill her. As they drove and the men became more and more incensed, the young woman, knowing she had little chance of survival, gave up hope. It was then she became conscious of her captives' intense need for her to show fear and to fight back. Instead she felt only pity for them and asked them repeatedly what was wrong with them and why they were so afraid.

The enraged young men could not cope with tenderness and compassion. They wanted a mirroring of their own isolation and terror. Instead she offered them a caring, safe place; the matrix denied them when they were children. The confused men abandoned their hostage by a road and sped off.

Is it possible then that empathy for others can hold its own over power and control?

## 6. FORCE VS EMPATHY

Upon my return to London from Europe, I was invited by a friend, a Welch guardsman, to a ball at his mess. A week after the ball I was invited to a dinner by my naval officer, who was still inexplicably in my life. This turned out to be the second occasion when he showed his sudden and unpredictable violence.

After a special, charming dinner at a naval mansion in the country, he dismissed the young man serving us, and locked the door. Another interrogation began in which he stated that there was no ball at the Guard's mess that night. Confused and mystified, I tried to tell him that despite not being in a committed relationship with him I always told him truthfully about my activities. He then asked if there was a lucky door prize that night which astonished me as it seemed irrelevant. The door prize seemed to indicate to him, however, that I was a liar and that the event I had attended was in fact a "Draw", a term I had never heard applied to an event.

As his anger became more intense, he continued to interrogate me with an attitude of entitlement that came from a life of command and authority. The more I answered honestly and without hesitation or guilt, the more furious he became. Suddenly his restraint left him completely and he flung his hands around my neck. His eyes were ablaze with fury. He tightened his grip on my neck as he shouted and raved.

In what now seems unfathomable innocence, I refused to enter his violent mind-space. I was aware of myself watching, spellbound by the remarkable reversal of character this man could produce. It seemed to me abysmally tragic that this wonderful man could have become so mentally unbalanced. It was as if sadness, disbelief and grief held me at a distance from fear. I could not enter his reality; my brain's frequency was giving me an entirely different perceptual experience from his.

I believe now it was my compassion for his plight that may have saved my life that night. On the way home, he once again became out of control with rage and raised his fist to thump me. I avoided his blow and, realizing finally that the man I had been so infatuated with had gone forever, I attempted to escape by throwing myself out of the car onto Bayswater Road.

He managed to drag me back to the car but by that time any remaining enchantment with him had been shattered completely. Apparently through a car accident that had caused him serious head injury weeks before my arrival in England. These injuries had so magnified the qualities of dominance and authority that had made him a naval leader, they now overruled his judgement as a man. He would never receive his greatest goal, the command of a ship and would never again be in complete command of himself.

Shattered by the experience, I quickly organised for my father in Australia to send money to convert my cheaper but longer return fare home by ship into a speedy return by air to the safe embrace of my family.

One could be excused for asking if patriarchal man had, in fact, become the demonic ape. Jane Goodall once said she

## Force vs Empathy

thought that chimpanzees were just like humans, only nicer. To her horror she was to find that her beloved chimps were more like us than she had ever imagined. As a former group of allied males separated and became a separate troop, the Gombe group set out on a search and destroy mission to kill their former friends.

One of my great passions is studying the great apes. Jane Goodall, Birute Galdakis and Dianne Fossey are heroines of mine whose lives among the great apes have inspired me and filled me with awe. It was not at all surprising, after reading the intimacies of social life among these incredible creatures, to find that the great apes share over 97% of our DNA. Tender bonding moments among chimp groups, however, are interspersed with violence and genocide that in the animal kingdom can only be compared in abhorrence with our own.

How often I have wished that the gentle Bonobo, the so-called pygmy chimp, also sharing 97% of our DNA, was closer to us in ancestry than their cousins, the chimpanzees. The bonobos have the most matriarchal, female-centric societies of all apes and they are also the most peaceful, cooperative and intelligent. Rather than resort to outright violence as chimpanzees so often do, they instead consider the welfare of the group as a whole and set about grooming and calming an upset companion.

These animals, standing upright and with their hair parted bewitchingly in the middle, make love not war, even settling differences this way. Even baby bonobos have been observed showing concern for a friend; one I remember was comforting a youngster who was fretting after being disciplined by the mother of another baby with whom he was becoming too rough.

Bonobos are perhaps living examples of life during humanity's matriarchal state, the time when humans lived cooperatively and selflessly together. I must admit to wondering, when sometimes observing the better side of human behaviour, whether there are the odd Bonobo descendants among the more noticeable chimpanzee-descendants among us.

Charles Darwin after giving examples of animals putting their own lives at risk for a friend, says, "For my own part I would as soon be descended from that heroic little monkey who braved his dreaded enemy [a large angry baboon] in order to save the life of his keeper, or from that old baboon who descending from the mountains carried away in triumph his young comrade from a crowd of astonished dogs - as from a savage who delights to torture his enemies, offers up bloody sacrifices, practices infanticide without remorse, treats his wives like slaves, knows no decency and is haunted by the grossest superstitions" [1]

Sympathy for animals lower than man, according to Darwin, seems to be one of the human race's latest moral acquisitions. It was apparently unfelt by the primitive tribes he studied, except towards their pets. The highest possible stage in moral culture though, he says, is when we recognize that we ought to control our thoughts. [I read this as their intellectual mind] Whatever makes any bad action familiar renders its reoccurrence so much easier. As Marcus Aurelius long ago said, "Such as are thy habitual thoughts, such also will be the character of their mind; for thy soul is dyed by the thoughts" Incidentally, Marcus Aurelius was born AD 121! [2]

As mentioned in the previous chapter, in order for the human race to develop fully to our next evolutionary stage, we first must change the habitual ways of thinking, the assumptions and beliefs that are leading us along devolutionary lines. We must develop the 'higher values" that Darwin spoke about in "Descent of Man", the capacity for unconditional selflessness and cooperation, placing the maintenance of the group and the environment in which we live above the selfish needs of the individual.

If our less advanced ape relatives can, under certain circumstances, achieve higher values, surely with our superior brain functioning we are capable of sustaining this behaviour. It is soul values which most benefited the peace-loving societies from which we descended, and it is these values that can bring us back into balance.

Using force and domination will bring only destruction and war and ecological devastation. War never ends, instead it becomes the reason for the invention of ever more cruel weaponry, it feeds on itself, and it is the means by which the vanquished return to conquer their enemies. A people who dominate cannot thrive, they must always remain vigilant and in "war mode", just a people subjected cannot grow and flourish as they are constantly in a mode of revenge. Power and control are not the means that will bring our salvation. Much the opposite is the case. Power and domination are frequencies of a lower order; they instigate change but cannot bring about transformation.

The ego, rather than the heart, is the foundation upon which power builds along with its allies, domination and arrogance. It would do the egos of the hubris of the collective human race a lot of good to learn that a good percentage of our 30,000 genes is shared with some of the world's most primitive creatures. Cultures with close relations to nature do not exploit the Earth on which they co-exist or try to dominate or control it. Instead they live in cooperation and harmony, with equality, with an appreciation for life and the interconnectedness of all that lives in the natural world. The threads of ancient wisdom are intertwined throughout the empathic society.

The question becomes, does man have a mandate to dominate the earth and every creature on it?

## 7. MANKIND'S RIGHT TO DOMINATE?

As a very long-term primate enthusiast, I made two trips to Taronga Park Zoo while in Sydney a few years ago. I sat comfortably in a shaded seating area watching established chimpanzee and gorilla families. These over time have become thoroughly habituated to human visitors and so lived as naturally as Jane Goodall and Dianne Fossey would have observed only after many years of adjusting them to the presence of humans in their sweltering and difficult jungle terrain. Previously when I had observed the chimps at the zoo, crowds had flocked around the fences and the animals ignored them completely, absorbed as they were in the intricacies of chimpanzee business.

I was privileged to be alone in the cave area on one occasion when an adolescent chimpanzee entered from her side of the shelter. I was putting my sunglasses into my handbag when the youngster bounded to within centimeters of me and sat on her side of the ledge with only a pane of glass between us. She watched fascinated and tried to reach for the bag to investigate for herself. I showed her the contents of my handbag and demonstrated their uses while she eagerly tried to join in.

Once my demonstration was complete, she began to dance in a circle, spinning around and around in moves I had never observed before, stopping only to come across to see how I enjoyed the performance before starting again. At the end came a truly magical moment. She came up to me and put her so-human hand gently on the glass where I also placed mine. Then she put her cheek on the glass where I met it with my own. For a few precious moments we communicated together, members of different species enjoying the close contact with one another. It was as if two million years of diverse evolution had never occurred and we were two sentient beings absolutely equal, accepting and as one. The magic of those moments, I will never forget.

Years before, during a whale watch in Ballina in Northern NSW, I noticed a humpback swimming very close to the breakwater. Anticipating that a second whale following behind would take that exact path as if following some invisible footprint, I raced to the end of the breakwater, arriving breathless just in time to see the great whale rise and blow within a few metres of where I stood. For what was probably only a second or two but seemed an eternal moment, the whale's huge eye met mine.

We gazed deeply, eye to eye, caught in an inter-species communion that was both intense and timeless. I stood in awe at the size and grandeur of this animal whose eye revealed a knowing recognition of me as another sentient being, small and comparatively weak as I was, but nevertheless one who shared that recognition. For a few larger-than-life moments it was as if two souls had met and acknowledged one another's beingness.

Dr. Tony Rose, conservation psychologist Anthony L. Rose studied and advanced the synergy of humanity and nature. Dr. Rose established the Epiphany Project (1994), the Bushmeat Project (1996), and the Wildlife Protectors Fund (1999). He compared the impact of interspecies communication to near-death experiences in intensity. One is never the same again. There is a meeting of kindred spirits recognizing each other as fellow beings following their own destinies on the same planet.

The conviction that the world was created for the sake of man found fertile ground in the Judeo-Christian tradition, where humans were given dominion over the earth and all living things. Many people still believe that human superiority is self-evident; after all we control the planet. This patriarchal argument has been used to justify the slaughter of noble animals as trophies for walls, and extends to the so-called superiority of men over women, whites over blacks, western civilization over native peoples throughout the world and the use of the environment as a resource.

It is this attitude that has wrought havoc on this planet. With patriarchy comes the sense of entitlement and superiority that has been responsible for the raping of the earth and endangering if not nearly destroying many of its species, including our own. These attitudes are both insatiable and unsustainable and based on the false premise of our entitlement. They are based on arrogance and greed.

It is our moral and ethical values such as those that assume patriarchy as the natural state which need to be reformed if we are to survive on this planet. Darwin explained last century that our moral progress will not be complete until we extend our compassion to people of all races, then to "the imbecile, the maimed and other useless members of society,"- and finally to the members of all species.

Students of animal behaviour have been accused of anthropomorphism because they observe characteristics, especially in primates, dolphins and the animals that choose to spend their lives in our homes, traits that appear to resemble human behaviour in many ways. These observations severely challenged those whose self-justification as rulers of the earth ultimately depended on their being the sovereign species. We now know that apes and other primates belong to the Anthropoid order which includes both ape and human, so it is our hubris rather than science that is responding in horror to any suggestion that mankind is not innately superior.

Rene Descartes, an influential thinker of the Seventeenth Century, in effect, disconnected mankind from the natural world altogether. Western thought based on Plato, considered that humans and only humans, were capable of rational thought. Evolution theory put a spanner in the concept of our uniqueness, but not in our ultimate conviction of superior mental capacity which somehow authorized us to dominate all species and nature itself. Studies of animals since that time have uncovered more rather than less human-like attributes than expected. As a result many observers became more attached to their subjects than was intended.

Darwin in the 18th Century proposed that the great apes, from whom we were descended, had the ability to reason and use tools, to employ memory and learning - all rational facilities. He even went further in a later book, "The Expression of Emotions in Animals and Man", by demonstrating instances of such as grief, joy, fear, loyalty and jealousy in many species, emotions that were previously considered to be exclusively human. Although this caused dissention at the time, few of us who have had the privilege of viewing David Attenborough and other animal specialists' documentaries can dispute that elephants express grief and many animals experience joy and loyalty to other group members to whom they are allied.

Birute Galdakis worked for decades in the forests of Indonesia with orphaned orangutan infants that had been psychically damaged as a result of poaching by humans. Aware as she was of the distinctions between humans and their ape cousins she nevertheless wrote, "The distinction between humans and orangutans had begun to blur in my mind. I had lost that gut feeling of separation, which is an integral part of Western intellectual consciousness. When orangutans are a natural part of the landscape, and your daily companions, it is difficult not to think of them as equals."[1] And yet we use caged and isolated apes and monkeys for scientific research as if they were objects rather than sentient beings much like ourselves.

As she worked among the orphans, Galdakis observed that orangutans display an honesty and candour that humans and chimpanzees cannot equal. "In their social machinations, chimpanzees remind us of ourselves. In their innocence, orangutans remind us of the Garden of Eden we left behind." [2] She adds, "Looking into the calm unblinking eyes of an orangutan we see as through a series of mirrors, not only the image of our own creation but also a reflection of our own souls and an Eden that once was ours. And on occasion fleetingly just for a nanosecond, but with an intensity that is shocking in its profoundness, we recognize that there is no separation between ourselves and nature." [3]

In one experiment, rhesus macaque monkeys were given metal "mothers", one equipped with a bottle, one with a cloth, 'in order to prove the importance of mother love', an experiment that cruelly exploited animals, deprived them of their mother's love, all in an endeavour to prove what all mothers, animal and human, know instinctively. The sacrifice of hundreds of monkeys and apes, subjects for research for the sake of science, has become almost routine. They are seen as objects, not as living creatures with feelings and emotions. Few people object to the psychological and physical pain inflicted on these ill-fated animals, justifying it from the point of view of benefits to humans rather than the viewpoint of the animals themselves who must have viewed their cages as torture chambers.

"When we turn our backs on compassion because compassion is too difficult and its execution too complex," Galdakis adds, "we risk our own humanity. We turn our backs once again on Eden and the state of grace given us there by the unseen hand of God."[4]

Dian Fossey, who worked intimately with gorillas just as Birute Galdakis did with her orangutans, likewise did not accept the view that humankind was superior to or should have dominion over the earth and its creatures. Instead she developed deep respect for these animals whom she saw as kindred spirits.

An extraordinary moment with a favourite gorilla, a huge silverback she called "Digit" demonstrates Fossey's attitude. While being filmed, Dian sat on the grass and busied herself with gorilla-like activities and grunts. Digit, not far away, beat his chest a few times before coming to less than a foot from her and sitting beside her. With giant hands he delicately picked up one of Dian's gloves and scrutinized it, turning it over and returning it to her. He took her note book and pen, gently returning both after inspecting them. Digit was enormous but his shy curiosity and gentle respect made it difficult to tell who the observer was and who the subject. Dian said it was as if millions of years that separate the two species had melted away. Digit proceeded to pay her the ultimate compliment by turning over and, showing complete trust, falling asleep.

This wonderful curious, trusting and gentle giant of an animal was eventually mutilated by poachers, an event from which Dian never fully recovered.

Roger Fouks, now in Vancouver, has worked with chimpanzees over many years. In "Next of Kin", he speaks of Washoe who in 1969 was a three year old female chimp to whom he was teaching American Sign Language. He was fascinated by the similarities between Washoe and his 2 year old son, Josh. In the morning Washoe would greet him with eager commands such as "Roger, hurry!", "come hug", feed me", "gimme clothes", Please out", "open door" which he says were gestural versions of what he heard from Josh every morning. Washoe would play-fight, and if she accidentally scratched Roger she would watch the blood and sign "hurt, hurt!" and "sorry, sorry" in exactly the same way Josh would when Roger and his games became a little out of hand. Frequently while dealing with Washoe, Roger would forget she was not a human. "She also understood signs, used them relevant to the situation, and the combining of symbols in an order that conveys meaning not nonsense in exactly what linguists define as syntax", he said.

All other emotions having been found to be shared alike by animals and humans, the one trait believed to finally distinguish humans from animals was the unflattering one of deceit. But even here we humans are not alone. One day Washoe tried to put one over Roger by using deceit. Washoe had been potty trained and knew she was never to enter the dwelling but to stay in her own domain, a very luxurious home of her own. Roger, on entering the house, found a deposit on the floor. Knowing Washoe had had an accident, he asked her who was responsible for the "dirty dirty", her own word for the offensive item.

Washoe, without hesitation, accused first his new assistant, Sue, then Roger himself, before eventually owning up in a display of repentance and need for forgiveness. "Washoe's dirty-dirty!" She eventually confessed following up with "sorry, sorry Roger" and "Quick, come hug", a complex range of emotional responses remarkably similar to humans of Washoe's age caught out in a forbidden act. [5]

Fascinated continually by the similarities between humans and orangutans, Birute Galdakis says it took her own child, Binti, to eventually remind her of the deep evolutionary differences. Binti, at two, was faster at everything than Princess, a three year old orphaned orangutan. Whenever Birute and her husband, Rod were working or an older orangutan was carrying out some operation or experiment of his or her own, Binti would watch, fascinated. To Binti, objects existed to be examined, manipulated and tested while the same tasks were foreign to the orangutan and had to be taught.

All developmental tasks orangutans performed, Binti was able to perform sooner, faster, and better. Once Binti reached two years of age the contrast between Binti and Princess, who was only a little older, became obvious, despite the fact that the two had previously almost been considered twins. For instance, although no one actually taught Binti signs, he learned not only the signs that another orangutan, Princess knew but also signs that her assistant, Gary Shapiro was trying to teach to Princess but which Princess refused to learn. Before even his first birthday, Binti showed signs of traits distinctively human; bipedal locomotion, speech, tool use and food sharing. Orangutans are capable of all of these behaviours but only at a later age.

Galdakis points out however that orangutans never develop these behaviours as fully as humans because they do not need them in the wild. Galdakis had to use these observations to maintain her awareness that her beloved orangutans were a remarkably similar but different evolutionary development from a shared ancestor. Their development was particularly suited to the environment in which they lived, just as Binti's was to his.

Unlike Binti who was fascinated with tools and utensils, aside from an ex-captive orangutan, Sugito who possessed the frustrating ability of disembowelling the camp's only electric generator, only rarely did Galdakis observe tool use among the orangutans. We must not forget though, that the solitary orangutans differ from other great apes in having little use for tools, complicated language and living primarily in the canopy of trees, of bipedal locomotion. Had these skills been an evolutionary requirement, these animals' propensity for learning in artificial circumstances has proven that they would have mastered them over time in the same way our own ancestors did.

Darwin said, "The difference in mind between man and the higher animals, great as it is, certainly is one of degree and not of kind. We have seen that the senses and intuitions, the various emotions and faculties such as love, memory, attention, curiosity, imitation, reason, of which man boasts may be found in an incipient or even sometimes in a well-developed condition, in the lower animals. They are also capable of some inherited improvement as we see in the domestic dog compared with the wolf." [6]

It is a horrifying fact that many great apes are the innocent victims of human exploitation, hunting for "bush food", population growth resulting in habitat destruction, and regional warfare as well as human greed which is the basis of our global economy. Dian Fossey, Birute Galdakis and Jane Goodall all felt that the line between humans, great apes, and other animals was tenuous and that the unstated law giving all rights to humans and none to animals is artificial and unjust.

None of them accept the view that humankind has the right to dominion over the earth and all its creatures. These women saw the apes as a kindred species, fellow citizens of the planet Earth. On the contrary, they felt that having promoted ourselves to masters of the earth, humans owe animals respect and protection. Our superior brain brings with it the responsibility to act as guardians and caretakers.

## 8. KINSHIP WITH ALL LIFE

Having rafted down the rapids from near Pokara outside Kathmandu to a wildlife retreat in the lowland jungle, I took the first opportunity to sneak away from the huts and spy on the private lives of the elephants. I marvelled at the intimate relationship that had developed between man and animal as I watched the mahouts wrapping elephant food in banana leaves and gently hand-feeding their elephants. After the meal the mahouts took their elephants for a swim, a treat obviously enjoyed by both man and animal as they romped together and sprayed one another with water from hand and trunk.

Later as we rode the elephants through the jungle with our legs spread wide over the elephant's flanks and swaying in time with the long strides, I absent-mindedly dropped the lens hood from my camera. Aware of the fact that any predator lurking nearby would not attempt to confront an animal as formidable as the elephant but would possibly not feel the same reluctance if I were rummaging on the forest floor alone, I made a swift decision that my lens hood should remain on the jungle floor as a dubious legacy of my visit.

The mahout, having become aware of my dilemma, began conversing in Nepali language with his elephant apparently concerning my lens hood. Astonished I watched as the elephant obediently explored the thick undergrowth and at the direction of his mahout, delicately lifted his trunk high and handed the tiny lens hood back to me.

Laurence Anthony, author of three books including "The Elephant Whisperer" was a legend in South Africa having rescued wildlife and rehabilitated elephants from horrific treatment by humans in many countries. Two days after his death in March, 2012 twenty elephants arrived from separate wild herds after having walked for days to his house. Somehow the elephants had intuited that Laurence, whom they trusted as a friend, had died.

It was obvious to Laurence's family that the animals, who had not been to the house for 15 months, wanted to pay their last respects as they do to members of their own kind. The elephants stayed two days and nights before making the long journey back to their own land. One year later to the day, the herd returned.

In "The Kingdom of Gorillas", Bill Weber and Amy Vedder say, "The world will be a much better place the day we learn to treat other species with respect and recognize that they, too, have certain rights. We have a well-developed notion of human rights that has evolved over time and is embodied in the Global Charter on Human Rights, a document signed by every member nation of the United Nations. Yet human rights are abused to varying degrees in every one of those member states. The rights of children, women, and national minorities are ignored or trampled on a regular basis. . . . We have so far to go in recognizing and respecting *human* rights that it is difficult to be hopeful for the adoption of *animal* rights any time soon." [my italics] [1]

As we shall discover later, David Hawkins, author of many books including "Power Vs Force", has calibrated various levels of human consciousness with a level of 200 being that of integrity, a virtue mankind has not long ago achieved and certainly not all of mankind has yet reached. Hawkins also calculated the level of consciousness of a female gorilla called Koko, who has worked for years with psychologists at the Primate Research Institute. Koko is a truthful, affectionate, intelligent and trustworthy ape who is capable of sophisticated sign language.

At one stage Koko signed that she wanted a baby! The stunned keepers, unable to produce a silverback at short notice, were quick to respond when Koko then asked for her second choice, a kitten. The kitten given to Koko was cared for as if it was her own child, with tenderness and consideration for its needs. Hawkins notes that "her [Koko's] integrity calibrates at 250. Thus one is safer with the ape Koko than with 85% of the humans on the planet." [2]

In the words of Darwin "Sympathy beyond the confines of man, such as humanity to the lower animals, seems to be one of the latest moral acquisitions. This virtue often of the noblest with which man is endowed seems to arise incidentally from our sympathies becoming more tender and more widely diffused until they are extended to all sentient beings. As soon as this virtue is honoured and practiced by some few men it spreads through instruction and example to the young and eventually becomes incorporated in public opinion."[3] Our acquisition of virtue however, is moving at a somewhat slower rate than one would wish.

Human beings, detached from the noblest values and with our trend toward power and control, regularly violate every facet of nature's design. In our arrogance, for instance, we consider the bugs that feed on our plants, an "error" in need of our correction and so inflict a host of deadly chemicals on the earth in order to correct that error.

Had we been in tune with nature rather than attempting to control it, we might have found more harmonious ways of limiting a preponderance of one type of bug. By refraining from over-hunting and poaching, for instance, nature's balance might have remained unaltered and the bugs kept in check by natural predation.

It has recently been discovered that the poaching of female adult elephants for "sport" [and for trophies celebrating the manliness of the hunter], actually causes trauma to the baby elephants who suffer something very similar to post traumatic stress disorder. Adolescent male elephants, deprived of training by their mothers in acceptable elephant behaviour, become unruly as they reach puberty and are now attacking humans, particularly those who are vividly remembered by the elephant as the perpetrators of their early trauma.

Rather than slaughtering all adolescent "rogue" elephants, a deterrent once considered the only option, rangers are introducing adult female elephants into the adolescent male groups to enculturate them into more acceptable behaviours. By resorting to nature's design, the rangers found that the rogue youngsters became calmer and less aggressive within a short period.

We must free ourselves from our hubris and accept that nature, free of our interference, is a network designed to work in balance and harmony. The world will be a much better place the day we learn to treat other species and each other with respect and recognize that animals as well as humans have certain rights, just by virtue of being cohabitants on our planet. Some people advocate the extension of rights to all living beings based on their capacity to feel pain and to suffer. Others prefer to start with creatures such as dolphins, whales and primates because of their higher intelligence.

I am proposing that we develop altruism as a means toward our own improvement as a species so that we can evolve to a higher level, just as "higher agencies" have done in the past. Altruism and compassion are the "higher agency" developments which according to Darwin and many others including Joseph Chilton Pearce, enable us to evolve as we are designed to do.

Richard Dawkins is an evolutionary biologist specializing in animal behaviour and a writer. He was the University of Oxford's Professor for Public Understanding of Science from 1995 until 2008 and at the time I write this, is now an emeritus fellow of New College, Oxford. Dawkins is said to have remarked that only human beings make the "error of altruism". According to this mind-set, human beings carry out unselfish acts only out of misguided emotion, such as guilt, and animals only demonstrate altruistic behaviour when caring for their young, or very occasionally when living in a large pack or herd. Many scientists believe there is no such thing as kindness in the animal kingdom, perhaps to excuse humanity's lack of it in the past.

In a study, carried out in 2006, Felix Warneken, of the Max Planck Institute for Evolutionary Anthropology in Germany, psychologists designed an experiment to test the existence of altruism where there was no expectation of reward and no intended evolutionary advantage, among both chimps and 18-month-old babies. The expectation was that if there was nothing in it for them, neither chimp nor toddlers would hand a stick to a stranger in order for them to reach food. In fact 12 of the 18 chimps and 16 of the 18 toddlers did just that. Altruism without any personal advantage is in fact, built into our blueprint.

Other evidence, from the Leverhulme Centre for Human Evolutionary Studies at the University of Cambridge, suggests that animal bonds are not always made for personal advantage. Studies previously demonstrated that chimps would affiliate with larger more aggressive individuals in the hope of an alliance giving future protection for themselves. This study, on the other hand, reveals that in many instances, chimps in conflict situations would ally themselves with the weaker individual rather than the bully especially if they were related.

Darwin gives an example reported in "Thierleben" 1864, of another biologist, named Brehm. "In Abyssinia, Brehm encountered a great troop of baboons who were crossing a valley; some had already ascended the opposite mountain and some were still in the valley; the latter were attacked by the dogs but the old males immediately hurried down from the rocks and with mouths widely opened roared so fearfully that the dogs quickly drew back.

They were again encouraged to the attack; but by this time all the baboons had re-ascended the heights excepting a young one about six months old who loudly calling for aid, climbed on a block of rock and was surrounded. Now one of the largest males, a true hero, came down again from the mountain slowly went to the young one, coaxed him and triumphantly led him away - the dogs being too much astonished to make an attack." [4]

Darwin gives many examples of animals behaving altruistically- dogs, even elephants, who deliberately protect people. "I will give only one other instance of sympathetic and heroic conduct in the case of a little American monkey. Several years ago a keeper at the Zoological Gardens showed me some deep and scarcely healed wounds on his neck inflicted by a fierce baboon. A "little American monkey who was a warm friend of this keeper . . . was dreadfully afraid of the great baboon. Nevertheless as soon as he saw his friend in peril he rushed to the rescue and screams and bites so distracted the baboon that the man was able to escape, as the surgeon thought running great risk of his life." [5]

If altruism is natural for some great apes, monkeys and many other animals, surely one can expect it of human beings. Any step toward taking the place in nature that our superior brain implies and protecting rather than destroying her species, is in fact a step toward evolving ourselves and saving our planet. But first we need to change our thoughts, our mind-set which dictates that we are superior and begin to see ourselves as part of a vast network that is life itself.

Animals have been perceived by humans over time as dumb, but time and again this notion has been proven to be false. Grizzly bears and elephants can recognize members of human hunting groups who have killed members of their families and have been known to wreak revenge on those particular humans while respecting others.

Some large groups of African elephants have become a "problem" because they attack certain native dwellings, the ones involved in interfering in elephant matters. Meanwhile a small family within this same elephant group, whose migratory path has been taken over by humans for a resort, the Mfuwe Lodge in Zambia, feel such trust in the visitors who respect and delight in the presence of elephants, that they comfortably walk through the foyer to browse on their favourite mango tree in the courtyard and even bring their new babies to show to the visitors.

We can't love animals or save wildlife, however, without understanding the social, economic and political context in which conservation occurs. "Our presence is perceptibly disabling the planet like a disease." says James Lovelock. He is referring in particular to our current thinking about values in economics dominated by deluded thinking about the supremacy of market forces and in government where public interests and those of the environment are ignored in favour of profit. "Rarely do we measure costs correctly" he says "thus the mess of current energy and transport policy and the failure to assess the likely impacts of climate change.[6]

We need to expand our consciousness to include nature and all its inhabitants and to understand the plight of animals and other living creatures, not anthropomorphically through our own eyes, but through the eyes of the animals themselves, just as Goodall, Fossey and Galdakis did. We must broaden our compassion and altruism, our "higher agencies", before we bring many species to extinction and destroy the environment that supports them, and this means reassessing our political, economic and technological values. Just as importantly, until we are prepared to accept the interrelatedness of all life we will remain unable to take the next step in our own evolutionary history but instead continue on our habitual path destroying and pillaging the earth.

The higher agencies including compassion and altruism are soul qualities that can only be attained when we let go of our former identity and the enculturated stories about ourselves that define, confine and limit us. Bill Plotkin says, "by deepening our identification with all life-forms, with ecosystems and with the planet herself we begin to discover within us what deep ecologist Arne Naess calls the ecological self or James Hillman calls a psyche the size of earth – the broader and deeper self that is a natural member in the more-than-human lifeforms." [7]

It is this ecological self that shifts our perspective toward a more humble, and humane way of being. The ecological self inspires a deep knowing that each of us is an integral member of an evolving world, a knowing based upon a sense of interconnectedness and interdependence with All That Is. It is in this way that we contribute to the evolution of consciousness.

Once our ego subsides, we can begin to really look into nature rather than just at it; to feel it, touch it, smell it and commune with it, free of any intention of harnessing and controlling it. Nature feeds the spirit, as all traditional peoples know. We so-called sophisticated humans are not schooled in reverence and awe. We pride ourselves on our technological prowess and rightly so, but in the process we have replaced reverence with a need to understand and analyse.

The problem with our scientific endeavours is not that they are innately wrong but that they are often lacking in ethical perspective and responsibility. Joanna Macy, eco-philosopher, renowned speaker and author of "The Great Turning" agrees. "To view the world as lover is to look at the world as a most intimate and gratifying partner . . . and participate in our world in a richer, more responsible and poignantly beautiful way." [8]

A solid basis for our ethical response in the world can only be gained when we humbly re-evaluate our place in an interrelated world. "The way to an ecological way of life", Thomas Moore says, "is to treat our houses as homes, our communities as homes, and nature as home. It is the intimacy in each relationship that serves the welfare of the other; at root, ecology is an erotic attitude of closeness, relatedness and care." He adds, "We wouldn't maltreat an ocean unless we had begun to think of it as a mere body of water and not as a spiritual entity. The Greeks and others who imagined the ocean as divine were not beneath us in sophistication, but ahead of us." [9]

Our disconnection from reverence in our everyday life is responsible for the utter insanity and inhumanity of mankind. Sacredness, on the other hand, is not something confined to a church or a special day of the week, but it is at the heart of our ongoing intimate relationship with our environment. It is what makes us truly human. This assertion is backed by the latest evidence from many disciplines, including biology, neuroscience and quantum physics which all state that life exists in a dynamic web of cooperation and connection. Lynne McTaggard sums it up when she says in "The Bond" [7] that nature's most basic impulse is not a struggle for dominion but a constant and irrepressible drive for wholeness and inclusiveness. Let us hope that history will prove this to be true as we stand at this crossroad in the lives of mankind.

### What Conclusions Can We Draw?

If our lives constantly influence and are being influenced by the environment, this bond demands that we relate to the earth in drastically different ways. Our subatomic makeup is indistinguishable from all matter in our world; there is no "us" and "other", only a constantly transforming "we". **We are basically living in tandem with our world in its entirety and this necessitates taking greater responsibility for our actions. We must choose co-operation rather than control and altruism over selfish gain so that we and our planet as a whole may evolve in a mutually sustaining manner.**

In a world irresistibly interconnected, we are being pulled by universal attraction toward a more direct personal involvement within Universal design. This is no random selection process. The accelerated growth in cortical brain functioning has allowed for an expansion of our consciousness. The higher energies required, however, are only accessible to those for whom the neural and psychological capacity becomes broadened. And this is a conscious choice.

So how does this relate to the direction human evolution should take? And just how do our neural pathways expand?

# 9. SACREDNESS WITHIN NATURE

When I reached the top of Mount Kosciusko, Australia's highest but hardly challenging mountain at 7,900feet, I knew I would find the others in our group, who were fitness fanatics, waiting for me. It was not that they would put me down for being so slow, after all I was an asthmatic, but I was making a real struggle of something which was really only a steady walk. I had an inkling that my problem was more an emotional than a physical one but couldn't quite put my finger on just what it was.

I wanted to be accepted by this group. They really enjoyed life and I needed a dose of that exuberance after a painful divorce. I had a few disadvantages though; they were seasoned skiers and I was a complete novice, so during the ski season I found myself at fairly regular intervals face downward, wiping snow from my goggles and, in a somewhat humiliating reflection on my own comparative skills, watching the littlies whiz by at breakneck speed, shamelessly flashing their "Thredbo Ski School Kinder" skiwear. Then too, there was the other little nagging recognition somewhere in the back of my mind; my companions had all attended private schools and lived in the Eastern suburbs of Sydney while I had attended a government school and lived on the less affluent South side.

Until my divorce, I had never felt "less than" others. It took the shock of my divorce coupled with my complete exclusion from my ex-husband's crowd, a group I had rather foolishly allowed to become my only social contacts during my marriage, to incubate a sense of low self-esteem within me. There had not been one word of one-upmanship from any of this ski group's members; they were simply absorbed in the walk and eager for me to appreciate it with them. My head trip was just that and nothing more. The only one who considered me "less than" was myself.

Once at the top, and I gazed at nature's wonder displayed below me, my mind-chatter quickly evaporated. A strong wind sprang up and we headed down. As we set out down the mountain, I noticed that all the mind-stuff that had given rise to my feelings of inferiority had gone. The others allowed me to lead, no doubt as a precaution against my being left behind.

We started out with the strong wind behind us. Rather than fight the wind, I leant into it, resting my weariness into it and allowing it to carry me. The slope assisted my legs to follow one another without effort. I felt a relationship with the trees as they too leant with the wind. The evening provided me with a coolness that was refreshing after the heat of the climb. The beauty and the silence entered my total awareness and every colour, shape, sound and touch of the wind, the smell of the pines, all became heightened and exaggerated. I felt utterly exhilarated.

I realised I was traveling very quickly but the speed had nothing to do with my volition. I was "being carried", and there was no temptation to think about how or why. I was flying, a part of everything around me, an intimate part. I was no better or worse than anything in my environment, instead it was a great privilege to be one with it.

When I reached the carpark, I realised the others were far behind, walking carefully and looking surprisingly tired. "What happened to you? They inquired, "You must have sprouted wings!" I had indeed.

Carlos Castaneda was instructed by Don Juan, a Yaqui Indian sorcerer, in using the "gait of power" when the body appeared to take over the reins from the mind, something that for Carlos, a Western trained anthropologist, was very challenging task. Don Juan explained that the gait of power occurred when the body and the world inter-related in what he called "energy from the inside/energy from the outside". By this he meant that our energy body interacts with the energy fields around it directly, and this interaction catapults us into a different experience of reality

In "The Teachings of Don Carlos", Sanchez comments, "The gait of power functions from the relationship of body/world. Don Juan calls this the entry into "left-side awareness" where you no longer find yourself as an ego voluntarily acting, rather it is your body that acts according to a direct relation it has with the world beyond the boundaries and narrow confines of reason and into the separate reality located beyond. [1]

So called 'widened perception' occurs when the world acts upon your body as opposed to our perceiving and acting on the world through the mental/emotional filter of the rational mind. The gait of power happens to you, it is not a conscious or voluntary action of the ego-self. It is as if there is total body awareness, a unity awareness in which we become a part of a whole universal consciousness. I believe that gait of power is what occurred while I was on Kosciusko.

Ed Mitchell in Kittyhawk had a similar and life-changing experience of expanded perception after his spacewalk. While staring out of the window, he experienced the strangest feeling he would ever have; a feeling of connectedness, as if all the planets and all the people of all time were attached by some invisible web. He could hardly breathe from the majesty of the moment. He felt distanced from his body as though someone else was performing his tasks of navigation. This too was an experience of unity consciousness.[2]

There is no way to separate the earth from its life; there is an exchange, an accommodation between them on a constant basis. An inter-relationship between the parts of systems combines forming a whole community of eco-systems. Scientists of today, spurred on by climate change, caused at least in part by our interference in the natural balance, are uncovering the intricate links in a system that is self-sustaining; between the primitive life forms and algae in the oceans, atmospheric chemistry, cloud physics, climate change and our need to accommodate and to adjust our behaviour to the requirements of the system as a whole.

All living beings act for their welfare by means of an interaction between the biological organism and the earth. The forest trees act as an enormous water pump, sucking water from the soil and releasing it by osmosis from the underside of its leaves into the atmosphere. The daily and nocturnal exchange of water, oxygen and carbon dioxide with the atmosphere helps maintain climate and temperature. The canopy protects the soil from the impact of the rain. With sunlight water, air and soil, the trees grow providing food and shelter to insects, the insects become food for birds, both of which pollinate and fertilize flowers and other plants.

Grasslands feed herbivores in turn provide food for carnivores. Herbivores eat fruit and distribute seeds in their droppings. All nature is an ecosystem, delicately balanced and self-generative. It does not appear by accident, nor does it contain errors that need correcting by mankind. The forest and its inhabitants provide food, energy sources, clothing, medicine and building materials for man but it is nevertheless a mistake to perceive it merely a resource for our use.

The interconnectedness of the whole Earth system, including its life forms has similarities with the quantum phenomena of 'entanglement' which states that when measured, one particle will behave in the same way as another particle in an energetic relationship that remains unaffected even by separation of large distances. This relatedness is a very different approach from the predominant scientific views of the Twentieth Century.

The accepted scientific view has been a very analytical and reductionist one based on Darwinian evolution and countless diverse disciplines, each confined to a tiny part of the whole. This reductionist approach does not view the world as a system, a web of interaction or an interconnected whole and despite their minute dissection and analysis do not stumble upon inter-relatedness, let alone Intelligent Design, at any point in their intense studies of life.

However dissenting voices such as biologist Kenneth Miller and theologian John Haught make a compelling case for the compatibility of Darwin and Divinity, and many renowned scientists cannot deny the existence of Intelligent Design within creation.

Einstein himself said,

"The more I learn of physics, the more I am drawn to metaphysics", and later,

"Everything is determined, every beginning and ending, by forces over which we have no control. It is determined for the insect as well as for the star. Human beings, vegetables or cosmic dust, we all dance to a mysterious tune, intoned in the distance by an invisible piper." And he concludes,

"The religion of the future will be a cosmic religion. It will transcend a personal god and avoid dogmas and theology."

James Lovelock in "Revenge of Gaia" says "we belong to the family of Gaia and are like a revolting teenager, intelligent and with great potential but far too greedy and selfish for our own good." Lovelock's theory of Gaia, a living breathing entity or earth was contested by Dawkins and others, but Lovelock insists the earth cannot be explained in classical reductionist terms. "It is time" he says, "that theologians shared with scientists their wonderful word, 'ineffable', a word that expresses the thought that God is immanent but unknowable." [3]

In "Descent of Man", Charles Darwin himself, the man behind evolution, says "The question whether there exists a creator and ruler of the universe has been answered in the affirmative by some of the highest intellects that have ever existed. If however we include under the term 'religion' the belief in unseen or spiritual agencies the case is wholly different for this belief seems to be universal with the less civilized races."[4] In our civilized world we consider ourselves far too sophisticated for belief in divinity within nature.

Paul Davies, a physicist, cosmologist and astro-biologist working at Arizona State University would appear to be a most unlikely person to write a book called "The Mind of God". In it he says "I do not subscribe to a conventional religion but nevertheless deny that the universe is a purposeless accident. The physical universe is put together with an ingenuity so astonishing that I cannot accept it merely as a brute fact . . . I have come to the point of view that mind – i.e. conscious awareness of the world – is not a meaningless and incidental quirk of nature but an absolutely fundamental facet of reality" [5]

Sir Arthur Eddington a Plumian Professor of Astronomy at Cambridge University, president of the Royal astronomical society, knighted for his scientific accomplishments, was without a doubt one of the world's leading scientists. He was also a mystic who drew the conclusion that physical reality is a creation of spiritual forces and intelligences. Sir James Jeans another leader in astrophysics also wrote of the existence of a non-material reality that stands in definite relation to the world of matter. The Nobel Prize winning physicist Max Planck, a famous German scientist says, "We must assume behind this force the existence of a conscious and intelligent mind. This mind is the matrix of all matter." [6]

Far from destroying the Divine within nature, science for the first time is proving the existence of an all-encompassing Intelligence by demonstrating that a higher consciousness is the bottom line of creation. There need no longer be two truths i.e. science and religion. If the same laws apply to the world at large as in the subatomic quantum world, we must conclude that we live within a matrix of indivisible and intelligent interrelation just as Ed Mitchell's experienced in outer space.

John Haught, theologian and Senior Research Fellow at the Woodstock Theological Centre at Georgetown University, speaks of an Intelligence that pours its creative essence into the universe and gives it free reign to go and make things happen; God voluntarily relinquishing control so new autonomous things may arise. He says it is through the random mutation and natural selection of evolution that Infinite Intelligence experiences its own potential.[7] This is intelligent design and evolutionary theory combined!

Bernard Haisch in "The God Theory" says "I have no problem accepting evolution, a billion year old universe, a big bang *and* a creator." [my italics] Haisch does not advocate

Intelligent design in the sense of "divinely micro-engineered life forms" being placed on the earth in complete and unchanging form, which he says would "seriously call into question the competence and benevolence of the designer". Complex biological mechanisms or properties of life-forms require so much complexity to fulfil their role that they could never have arisen one step at a time over eons.

"Indeed it seems to me" Haisch adds, "that there is better empirical evidence for the existence of God than there is for the many dimensions of string theory."[8] As an astrophysicist accustomed to observation as a scientific method for uncovering empirical evidence, he nevertheless could not dismiss the experiences of mystics throughout the ages of transcendent, supernatural realities.

The view of the perennial philosophy is that during vast expanses of time or "yugas", within each a single cycle in the evolution of God, there is an "inbreathing and outbreathing" of all creation; periods when all remains as fields of potentiality followed by periods of manifested reality. Professor William Tiller, Fellow to the American Academy for the Advancement of Science at Stanford University, explains that there are periods where all is in a state of perfect coherence [unmanifest] and others when there is an unravelling of the coherent state as life unfolds and lower frequencies manifest. Coherent God-energy then splits into individual souls that reincarnate in a multitude of embodiments as the soul journeys on its evolutionary path.[9]

It has been accepted among many theologians and thinkers that God became manifest so that Conscious being could experience doing in a relative world of polarity, that is, one which makes experience possible. Living forms are an expression of God, Tiller says, because physicality is the only way to know experientially what is already known conceptually. In the words of the American author Neale Donald Walsh, the realm of polarity, good and evil, is necessary because "You cannot experience yourself as what you are until you encounter what you are not." [10]

Deepak Chopra says something similar. "Reality itself may be only a symbol for the workings of God's mind, and in that case the "primitive" belief - found throughout the ancient and pagan world - that God exists in every blade of grass, every creature, and even the earth and sky, may contain the highest truth. Arriving at that truth is the purpose of spiritual life." [11]

Creation then, appears to be continually in progress rather than being a completed fact. It is, however, a process constrained and directed by the basic laws of a Higher Intelligence. There is a self-creative, self-organizing characteristic to living things suggestive of the influence of a higher-order Intelligence. It would seem an Infinite Intelligence is translating potential into actualization and as a part of that Infinite Intelligence's creation, we ourselves are the living, breathing actualization of God consciousness. As Bernard Haich puts it, "Every experience of every individual consciousness springs from Infinite Intelligence and will return having been transformed by experiencing the universe."[12] Human consciousness is designed by a higher or Divine Intelligence to evolve. The two are no longer in conflict, but rather are part of Creation in progress.

The French Jesuit palaeontologist, Teilhard de Chardin also proposed that evolution does occur but in a directional goal-driven way. He used the term "Omega Point' to describe an aim toward which consciousness evolves in an evolutionary process converging toward final unity.

If we perceive God's creation as a whole inter-related web, evolving over time within universal laws it is only one step further to recognizing ourselves as cohabiting within the whole Divine evolving living ecology - trees, animals, the earth itself. As Haisch says Infinite Consciousness, using the Big Bang as a tool for its own evolution and growth, creates and shapes the matter we see as reality and within which we participate with other conscious life forms.

Evolution would appear a pitiless indifferent process, Haisch believes, only if we fail to recognize ourselves, along with all of creation, as participants in God's very being. Far from existing in a dog eat dog survival battle, we are incredibly, God's expression in human form of Its/His/Her very being.

Now that little bombshell should demand a re-appraisal of our lives, if anything could!

If what all these authorities say is correct and our consciousness is part of the earth's unfolding and the evolution of human consciousness, it surely would behoove us to keep an awareness of ourselves within an interrelated, even sacred world. We could no longer feel separated once we become aware of focusing our attention on interconnectedness or context rather than the world as content, separate and different material objects and events.

## How Would We Behave If This Were the Case?

Thomas Moore, quoted in the last chapter, says "spirituality, nature and ethical behaviour go hand in hand. If anything, we have lost the one thing that would sustain our intimacy with nature - a religious sensitivity to the sacredness of all forms in nature. .... we could know this only if we were deeply schooled in the necessary virtue of reverence."[13] He calls this sensitivity to the sacred, "an enchantment with life" which arises when we move so deeply into an experience that "its interiority stirs the heart". Enchantment or reverence can be the basis for an ethical response to the world, to see it as our home where all beings are inter-related and are entitled to respect. We would not violate the earth if we perceived it as our sacred home.

To perceive the sacred is to be in love with life. To love is to hold our beloved close to our hearts.

A former archbishop of Canterbury, Rowan Williams is quoted as saying "the relation of God to creation . . . .is the relation of an external activity which – moment by moment- energises, makes real, and makes active what is there. I sometimes feel that a lot of our theology has lost that extraordinarily vivid or exhilarating sense of the world penetrated by Divine energy in classical theological terms." [14]

Gary Zukav is a spiritual teacher and author of "The Seat of the Soul" and "The Dancing Wu Li Masters" which won the American Book Award for Science in 1979. "Reverence" Zukav says, "is engaging in a form and a depth of contact with Life that is well beyond the shell of form and into essence." Within this essence, the beingness of all things is respected and process is honoured and revered. There is an ever-present consciousness of the unfolding, the cycles and the sacredness of life, of all things evolving simultaneously with our own evolution. There is a constant awareness of the interdependency of different species and "the significance of each creature to the compassionate unfolding of the Universe."[15]

This to my mind is the enchantment with nature that Thomas Moore speaks of, a real sense of how it feels to experience the sacredness within Nature.

Pagan peoples were perhaps ahead of us in this regard; they saw nature and spirituality as intimately linked. Native American plains peoples honoured the Great Spirit in all things. Chief Seattle of the Squamish people in what is now

Washington state, delivered a letter in response to the government's offer to buy the remains of Salish land in the 1800's, "every part of the earth is sacred to my people," he said "We are part of the earth and it is part of us." This letter is well worth quoting in more detail;

"If we sell you our land, you must remember that it is sacred.   What befalls the earth befalls all the sons of the earth. . .   This we know; the earth does not belong to man, man belongs to the earth. . . All things are connected like the blood that unites us all. Man did not weave the web of life, he is merely a strand in it . . . whatever he does to the web, he does to himself.   One thing we know; our God is also your God. The earth is precious to him and to harm the earth is to heap contempt on its creator. . . . . .   We love this earth as a newborn loves its mother's heartbeat. . . . So, if we sell you our land, love it as we have loved it."

Chief Seattle then asked the white men a question,

"Will you teach your children what we have taught our children?   That the earth is our mother?"   Chief Seattle requested of the white man that he "care for it as we have cared for it. Hold in your mind the memory of the land as it is when you receive it. Preserve the land for all children, and love it, as God loves us all. As we are part of the land, you too are part of the land. This earth is precious to us. It is also precious to you. One thing we know; there is only one God. No man, be he Red Man or White Man, can be apart. We are brothers after all." [16]

Have we taught our children that the earth is our mother? History has taught us the tragic answer to that question.

Traditional peoples know that co-habiting our world is a sacred privilege, that nature feeds the spirit and that we are poorer as a people for ignoring our spirit in favour of our rationality. So called "primitive peoples" honoured the earth and its creatures and so left no destructive footprint behind. How sad that we have not taught our children that the earth is our mother, that the land and its inhabitants in the centuries since then have been treated more as a resource rather than an entity to be revered. How tragic that our arrogance has prevented our seeing that mankind is not in control of the web of life but instead is just one strand within it.

Carlos Castaneda was taught by Don Yuan, "If we stop perceiving ourselves as egos, accepting instead that we are fields of energy, then not only will we have to change the way we view reality, our way of behaving in it will tend to change as well. As egos we feel compelled to defend and reaffirm an enormous number of actions in the name of ego; as fields of energy our attention must be placed on how we use that energy." [17]

To use our energy appropriately we must, in Don Yuan's words, behave "impeccably", that is, a true warrior's task is to experience our world in its totality and carry the self beyond the limits of "personal history"; we must jettison the stories we tell ourselves to bolster our identity and justify actions that could in no way be described as "impeccable".

With sensitivity to our place in nature, we can no longer view the world as inert but as alive and enlivening. Spiritual life is not enclosed within a church to be unlocked only on Sunday. Nature itself can be a portal into another way of being. As we walk in nature we can allow the mountains to reveal themselves powerfully to us, we can stop for a moment while a river sings its rippling song and the wind rustles the leaves. We can attune to the moods of the seasons, the joys of spring, the rustic colours of Autumn and the quiet waiting time of Winter. Then the seasons will stir our souls so that each change becomes an initiation.

Attunement with nature brings a different experience of reality just as my experiences at Mount Kosciusko gave me. Experiences such as these give us a brief glimpse of unity experience. They shift our perception of life and initiate us into the possibility of living with a new perspective. We experience briefly as a child experiences, as if for the first time, without an overlay of the past stories, without rationalizing or categorizing, but as a participant with nature. All that once what appeared to be separate becomes coherent, cooperative and supportive. Then, in that moment of intense whole-body focus, the heart begins to expand to include All That Is. With unity awareness, objects change in significance or as Sardello says "the world shifts from an 'it' to a 'thou' in a moment of creative, imaginal perception requiring presence to the radiating quality of the heart sense." [18]

This is Thomas Moore's "Enchantment". "Enchantment" to Moore involves a spirituality that is deeply rooted in nature and which has the power of ushering us into a way of life that is closer to the soul.[19] Sardello puts it so poetically, "through this sense, the individual soul is held within the world soul." [20]

We experience a unified world when we experience through the synthesizing energy of the heart. We then experience it as if for the first time with what the Buddhists call a sense of "not-knowing" – not categorised by our rational mind. Through the heart the world becomes an interconnected whole, each part is seen as a microcosm of the macrocosm, as if seeing "the world in a grain of sand, and Heaven in a wildflower", [Blake] a hologram in a world of interpenetrating realities. If we are part of a world soul we must become familiar with a new vision of the world as one.

What is experienced in unity consciousness is the harmony and union of one's being and nature, one's consciousness and one's conscience. We deeply appreciate the correlation between physical existence and spirituality in the depths of the heart and expand that physicality to include humanity as a whole. It then becomes imperative that we act in accordance with nature not against it.

Moment by moment we become aware of the manner in which we relate, and interact within All That Is. When we are isolated from our spiritual being we perceive only the familiar and mundane, lacking inspiration and enchantment. Lower levels of consciousness such as these hinder our progress; they are in fact devolutionary. In order to evolve and to really comprehend infinite natural connectedness at a deep level, we need to extend and deepen all our sensory faculties beyond those we use in everyday consciousness.

It takes a higher level of awareness, appreciation and attention to enable us to see and understand sacredness and divinity and their relationship to our heart and soul.

Philosopher Duane Elgin says, "I believe that the most far-reaching trend of our times is an emerging shift in our shared view of the universe—from thinking of it as dead to experiencing it as alive. In regarding the universe as alive and ourselves as continuously sustained within that aliveness, we see that we are intimately related to everything that exists. This insight ... represents a new way of looking at and relating to the world and overcomes the profound separation that has marked our lives.[21] In order to experience reverence, a particular awareness is required which honours the numen or divinity in all things. Divinity is not something set apart from life, but it is the very essence of every dimension of life.

The ancient wisdom of the Upanishads says "This universe is the creation of the supreme power meant for the benefit of all his creations. Individual species must therefore learn to enjoy its benefits by forming a part of the system in close relation with other species. Let not any one species encroach upon the other's rights" [22]

We will need a comprehensive change in consciousness in order to consciously and deliberately cooperate with the Divine intent, during this potential Turning Point in history. We must become what Plotkin calls "a soul centric person" to consciously experience in our everyday acts of living this interrelationship and interconnectedness with everything else.

Nature has wondrous ways of revealing its awesome design when we abandon our left-brained thinking for a moment and just observe. When we become too egocentric and fail to cooperate with nature, the living planet has a way of reminding us of the error of our ways. Sometimes the reminders come in the form of disastrous landslides, earthquakes, tsunamis or global warming. At other times Nature has other more gentle methods of reminding us if we care to become aware of their significance.

Freddy Silva, who has made an extensive study of crop circles in the United Kingdom, believes crop circles are one of nature's ways of nudging us to raise our vibrational rate so that we experience this understanding. The circles are composed of symbols based on universal principles all of which are based on geometric structures within nature and the human body. The crop circles' symbols are therefore capable of bypassing the brain's left hemisphere of reason, enabling the exchange of information to take place at a cellular level. This, he says, enables individuals if they so choose, to raise their vibratory rates and so prepare them to receive this language through the heart. Freddy Silva believes the crop circles are physical expressions of up-wellings of energy from the living, breathing planet that occur in order to remind us that we are part of a greater reality, that we are not egocentric but cosmocentric.[23]

We have models already from which we may learn ways of changing our perception and modes of action to aid us on our evolutionary path. In the US, Thomas Berry, the American naturalists John Muir and Henry David Thoreau all of helped inspire environmentalism in their own ways, guided by spiritual affinity with the natural world.

Organizations like Greenpeace and the Sierra Club, World Wildlife Fund and World Vision are addressing our need to redefine our values concerning commerce and its relation to the environment. In Australia, Bob Brown and the Greens, Peter Garrett and Tim Flannery and currently participants in Extinction Rebellion are among the many well-known campaigners who have fought against destruction of the environment on many issues.

Societies such as Medecins Sans Frontieres, WSPA and many other organisations throughout the world as well as many individuals including actors and other celebrities are working for protection of the environment and animal rights.

Nothing ignites that spirit more than disasters such as the bushfires that are raging through Australia at the time in which I write. It is evidenced by the outpourings of grief at the loss of life, human, animal and environmental.

Beside the destruction reside movements for change. Such is the dualistic nature of Divine evolution.

## 10. "CREATING YOUR OWN REALITY"

If, as many new age advocates suggest, we create our own reality, how is it that we have made such a mess of it?

Events we experienced at the end of our four year residence in San Francisco thrust my husband and me into a world where the rules we had always lived by were tossed on their heads. Ian had joined a business that at first appealed to our sense of charitable beliefs and the possibility of expanding the business back home to Australia.

To our astonishment however, the situation soon took us into a foreign existence where the words "scruples" and "honesty" became meaningless and people were mere pawns in the schemes of an unprincipled woman, Ian's boss. Soon we were to hear stories of her connections with criminal elements including the Mafia and reliable reports that she had parents who were both CIA agents. We also learnt of her previous use of both legal action based on false accusations, and threats against those who crossed her.

As events proceeded we felt we had somehow entered the set of an all-too-real horror movie. Never before had we believed in an evil energy or satanic force. Now we felt we were facing something truly resembling one head-on. Our beliefs were totally turned upside down. I was just 'the girl next door", someone considered the Pollyanna type while Ian was an idealist, an environmentalist and a peace lover. We were not even attracted to movies that depicted scenes such as those that appeared at this time in our lives.

Many times I have noted that when my clients and friends face adversity, it appeared in multiples rather than single events.  As if the events in the US were not enough, unbelievably events in Australia simultaneously deprived us of the home we had built together and loved.  The tenants in our Australian home, having failed to raise the funds necessary to buy it, informed our agent and our solicitor that if we did not agree to sell the place at a very much lower price that our belongings stored in the attic could disappear. They refused to pay rent as a further incentive.

We had met with two situations in two countries quite divorced from any experience we had ever known. Through no apparent fault of our own, we had been forced to leave our adopted home in California and to sell our house in Australia. Many of our treasured belongings were stolen in spite of our agreement to sell, and we had lost very nearly all of our savings invested in the US start-up company.

Our belief system and sense of security had been turned upside down. Ian and I were in a state of complete shock. The bottom had fallen out of our world.  The collapse of our financial situation had dragged with it all those parts of our egos which had been attached to security, trust and normality. Perhaps inevitably, the stress effects became a wedge, that drove Ian and I to separate as we negotiated our own individual ways through the shock and grief.

It was then I entered what I now see was a dark Night of the soul experience.  I joined a New Age group hoping to find some comfort and support and perhaps a little encouragement to help me wend my way through the distress.  Instead I was met with, "Why on earth did you create something like that in your life?" and "It's your attitude that brought this to you and now it's your attitude that is keeping you stuck!"

The New Age "create your own reality" movement holds that you manifest your life experiences with your state of mind and your attitudes. Positive thinking, they say, brings a life full of positives, a negative attitude brings your dramas and disasters down upon you. Creating your own reality, especially as it is portrayed in the movies, "The Secret" and "What the Bleep", promotes getting whatever you want by visualizing it clearly and setting your mind positively on that goal, be it attaining business promotion, passing an exam, possessing the biggest car or house of your dreams, or finding the best looking or wealthiest partner. My husband and I were told our "creation" of disaster in our lives was an indication of our own negative emotional state; in a word, we were not only homeless and penniless but we were also single-handedly responsible for all of our difficulties.

At the time this stance appeared cruel and uncaring, utterly lacking in empathy. Even now this mind-set, despite the fact that it might be loosely based on ancient beliefs, appears to me to have several spiritual and psychological shortcomings. Rather than any sense of greater purpose in life, it promotes using the mind for self-centred motives. As such, it bases its premises on the ego values of status, self-importance, living a carefree life and impressing others. These are hardly synonymous with the higher values we are speaking of when we speak of evolution of mankind.

Ken Wilber, the Integral psychologist, believes as I do, that thought does not in fact create reality; it merely interprets it. He speaks from personal experience about how people can become unintentionally cruel with their interpretation of "creating your own reality". Ken Wilber's own experience of misinterpretations causing unnecessary emotional pain occurred when his second wife, Treya, was dying of cancer in 1989. He and Treya had guilt trips laid on them by healthy people, friends and healers, about Treya's conscious creation of her illness. Comfortably sheltering behind their positive thinking, these people, like those in my spiritual group, were able to dissociate themselves from compassion with the belief that people are solely responsible for all that occurs in their lives.

This attitude can too easily step over the line into judgment and self-righteousness. Had they had a real sense of a higher Reality, these critics would be operating from their hearts rather than their minds. People's beliefs and interpretations are merely ideas about the truth, perceptions placed upon some observable outcome, whether it be illness or the receipt of a wonderful legacy. The truth behind the cause of any event, however, is based on a much larger inter-related field of factors, much of it far beyond our intellectual knowing, let alone control.

Richard Moss, medical doctor and teacher of the psychology of consciousness speaks of just such an attitude when he says, "much of the New Age scene represents a multitude of forms of egoic specialness and glamour, sophisticated defences against real social and interpersonal responsibility, against the deep humbling that life must inevitably deal." [1]

Real spiritual substance, according to ancient tradition, can only come by living deeply and allowing ourselves to experience the deep troughs that life presents along with the wonderful heights. Human beings must sometimes be brought to their knees in order to break down the powerful grip of the ego. To try to avoid this is to step onto a regressive path in order to prove one's spiritual superiority.

This type of New Age ego can too easily stand arrogantly outside its experience rather than consistently altering, adjusting and growing according to the experience. In so doing, it misses the important learning. Deeply experiencing the highs and lows of life reframes our viewpoint so that we evolve from adolescent narcissism to adult compassion, from cynical stances to caring responsibility, from self-involvement to consideration for the planet and all its inhabitants.

In my opinion, movies such as "The Secret" and "What the Bleep" have served to ignite the misinterpretation both of modern physics and ancient beliefs on which they are supposedly based. One look at the websites shows the level of consciousness from which they arise. The ideas behind the movies are loosely based on a quantum physics principle called "the observer effect" which they interpret to mean that it is our personal observation alone that turns quantum possibility into reality. Dealing as it does with the minutiae of sub-atomic waves and particles however, Quantum physics cannot easily be made to apply to the world at large.

The theories are then extrapolated further until the universe itself resembles a personal mail-order catalogue with which you select the actual experiences you would like to have and those you wish to avoid, along with obtaining a string of possessions that serve to reinforce your ego-status.

But the perennial philosophy tells us, this life is a theatre upon which you progress and evolve through all the vicissitudes of living. By attempting to design your own journey for your pleasure and comfort, you miss the point entirely. Spiritual growth is, after all, about moving toward a state of "being", a quality of detached awareness rather than of "doing" or "possessing", both of which serve to enhance our small sense of self.

In fact, Benjamin Libet, a pioneering scientist in the field of consciousness at Harvard University, found that choice actually happens outside of conscious awareness. Choice is not a single separate activity but is connected to everything else that is going on in us and around us. The brain, informed by our biochemistry, actually begins processing the desired outcome before we consciously "decide". Any conscious decision to act comes after our body has already begun the process of choosing. Many of our decisions are being made in the morphic field without our conscious knowledge of it. Mind is much larger than our individual ego believes or our self-absorption allows.

By teaching that the world quite literally revolves around you, "Creating Your Own Reality" beliefs serve to entrench us in ego states such as like and dislike, attraction and rejection, good and bad, want and don't want. "The Secret" teaches that you have the power to manifest a new car, cure your illness and create the wealth you want, but that is just not the way creation works. Our desire for a trouble-free and indulgent life flies in the face of <u>karmic laws which bring you what you need for your advancement</u> toward higher consciousness.

We evolve spiritually by experiencing the yin and the yang, the "good" and the "bad". Our third dimensional duality contains these polarities for a purpose; to aid us in evolving toward an awareness of being one, of attaining unity consciousness. But by trying to shape our own reality and attempting to control experience along our own preferred lines, we avoid true presence with what is actually occurring. As John Lennon once said, "Life is what happens when you're busy making other plans." You cannot be one with everything and feel unity with all that is, while at the same time concentrating on what the mind believes it wants i.e. a new car, more money, fame or status, or what it wishes to avoid-i.e. disease, grief and pain.

Richard Moss, psychiatrist and author believes many of us look at the cause and effect of creation in the crudest terms and therefore seek to place blame - on God, our faults or more likely  someone else's, misfortune, or even government policies.  Moss says, in truth when life wants to, it throws us a curved ball that faces us with our own issues, those getting in the way of our personal and spiritual growth.  Far from being a misfortune, it is just these events that have the power to transform our ego states.

It is impossible to really experience awe, reverence, compassion and any sense of the sacred, without having experienced their opposites, fear, profanity, cruelty and hatred.  Moss's view is that some people act as if they are members of a herd, their view is myopic and centred only on desires dictated by enculturation.

One of the concerns about our minds creating reality is that the human brain is composed of unconscious shadow elements that often interpret experience along distorted lines, incorrectly influencing our perception, our experience and our responses.  When these distortions coexist alongside our attempts to create our reality, the result can be toxic, engulfing us in the energetic dimension of "dog eat dog" narcissism and personal gain.

According to perennial philosophy, rather than controlling events with our minds, we should maintain awareness of what is happening without judgement and respond with wisdom, thus remaining in harmony with what IS.

Deepak Chopra agrees.  "What's much more dangerous here is solipsism, believing that only your mind is real, while all objects out there in the world are mirages that depend upon you, the perceiver, and without you they would melt away.....The ego, as it takes charge of holding the world together, forgets that creation depends upon Grace. We make our choices, some good for us, some bad and then Grace shapes the results."

In a later book he adds that your ego does not control your daily life. Events unfold unpredictably. We are not in charge of how things work out. Our responsibility lies in adjusting and growing as we find solutions to whatever occurs and with the awareness that it happened for a higher purpose. "Despite all the clichés about making your dreams come true," he says, "no one is really taught that success depends on your state of consciousness.[2]

It is from infinite potential or what Chopra calls the "field of pure potentiality" that "the manifest" or reality as we know it unfolds, or using Bohm's terms mentioned earlier, it is from the Implicit order or Super-consciousness that the Explicit or manifestation arises. Laws of creation govern manifestation and these laws involve vast inter-connected, nested, holographic dynamics, dynamics far beyond our brain's capacity to control or even understand.

According to David Hawkins there is no such process as causality. Universal Awareness operates automatically behind the machinations of our scheming minds. Hawkins explains, "If the unmanifest becomes manifest through continuous creation, then no other intellectual devices or premises are required as attempts to explain the obvious. Everything is created to be self-evolving because that is intrinsic to its existence and the nature of Creation itself. The Creator and created are identical." [3] All manifestation is inherent within the automatic process of potential actualizing which is in itself innate to the nature of the Creator.

It is easy to become confused by language when speaking of spirituality and of science in one sentence. Coupled with this difficulty, each writer and researcher uses different conceptual frameworks and of course those sages who report experiences of Oneness or Higher Consciousness use individual words to express the inexpressible. I am attempting here to use terms that as near as possible convey the interconnected ideas.

To put it another way, if we perceive 'b' occurring after 'a' on a continuous basis we assume that 'a' causes 'b'. What we fail to see is that AB is creating both 'a' and 'b' on a higher level. What we assume to be causes, such as our intentions, ideas, thoughts, and actions are merely means or tools in the physical by which potential actualizes.

Studies at the Institute of Noetic Sciences suggest that intention can "affect" or "weakly influence" living systems to some degree, but this is far from creating our day with our mind. The brain and its thoughts are not responsible for causation but merely for the observation and interpretation of what already is. The brain does not have the power to "cause" anything; its purpose is to work upon existing elements.

Hawkins gives the example of brain-dead people in hospitals who have experienced profound and transformative near-death experiences and who survive to report them. Their experience occurred entirely without the benefit of the brain's activity. As mentioned previously, Harvard trained neurosurgeon, Dr. Eben Alexander in his book, "Proof of Heaven", gives an account of his own near-death experience when his frontal cortex was not actually operational because of a meningitis-induced coma.

### What is the Nature of "You" and "Reality"?

The individual consciousness is made up of the ego, inner or Higher Self and Higher Consciousness but it is the ego that places itself in the prime position where it has no right to be. It is in this way that 'Thoughts influence reality' becomes 'Thoughts create reality'. The Higher Self that is one with Spirit and the Higher Consciousness that actualizes reality, have little to do with your ego self.

Reality is created by a multiplicity of factors on many planes not only the mental or physical. It is the Absolute, the Field of nonlinear intelligent energy or universal unity consciousness [God] that is the matrix, the primary source of all manifestation, differentiating itself into levels of linear form that actualize potentiality into matter. All that exists has a source but not a 'cause' as such, because Creation is a continuous, ongoing self-perpetuating process which is witnessed sequentially by mankind as evolution. Evolution expresses creation, not causality. Reality is an interconnected web where everything is influenced by everything else and in which there are no causes and effects as such but only continuous creation.

There is a natural pace to this continuous creation or universal unfolding. Rather than attempts at manipulating or demanding our own ego preferences and timing, we benefit more when we allow things to unfold and manifest in harmony with the universal whole.

British scientist and author, Rupert Sheldrake in 1981 said form or physical form occurs first within the field of consciousness with morphogenic patterns which produce a living, developing universe with its own inherent memory. It is "morphic resonance" within the universal frequencies that is essential to convert potentiality into actuality.

Current string theory postulates that ultimately all matter in the universe is composed of vibrating filaments [strings] or membranes [branes] of universal energy. Actualizing potential or transforming the unmanifest universal energy into the manifested physical state is possible only with the infinite power of this basic energetic substructure. We evolve when we align, not with the intentions originating in our brain, but with higher dimensional creativity processes. In this way we co-create within this universal dynamic.

In "A Dialogue on Science and Spirituality" in 1996, Rupert Sheldrake and Matthew Fox say, "Co-creation is still ongoing and the Great Spirit is still with us – and we don't control it. We call on it but it blows where it wills." Expanded mind [one with a more developed neuronal structure] can affect the field, and intention within the field can affect nature. They give the example of co-creation with the will of God in the form of prayer which takes positive thinking and desire into a conscious alignment with higher purpose.[4]

William A. Tiller, Professor Emeritus, Stanford University puts it this way; "It is the prephysical reality where we collectively create our future at the space-time level." This higher domain is within a frequency where there is no separation but a unity. The events that occur in this unseen domain are then the precursors to all the events that materialize at the space-time level of our daily reality.[5]

Consciousness theorists and physicists Edgar Mitchell, Peter Russell and Amit Goswami have made extensive studies of consciousness and manifestation. They believe divine intention activates physical manifestation in a kind of super-synergy. Russell and Goswami suggest that the Consciousness pervading the universe is immensely creative, that it animates and enlivens the material world, and because we are expressions of that consciousness, we have the opportunity to participate with it in manifesting events in our daily lives. The way we do this, they say, is not by mental power but by spiritual states of meditation and stillness, that is to say, in a higher state of consciousness.

"With a still mind, it becomes possible to perceive that our true nature is awareness itself, the field from which all phenomena arise and are sustained. We can consciously draw from this dimension of our nature that we share with all existence, direct it with intention, particularly when we are in a meditative state."[6] This takes the concept of synergy into the spiritual realm; an evolutionary leap from mental conjuring.

Mystical traditions speak similarly of a Source out of which arises all phenomena in a continuous evolution, the unfoldment or manifestation of which we observe and participate in as it progresses. It is from the Infinite or God, the one synonymous with All That Is, Universal energy or the God-force that all content unfolds as an act of Creation innate to the nature of that Godhead. God supports the created universe in an ongoing continuously sustaining act of creation.

Amrit Goswarmi, Indian philosopher, expresses the view shared by the great a majority of esoteric philosophers that consciousness is the ground of all being, the root cause of manifestation. His belief is that the ultimate purpose of humanity, as of most sentient beings, is to consciously participate in the expansion of consciousness. In their view consciousness is an irreducible feature of reality, the primary reality. Current findings in evolutionary biology, neurophysiology and quantum physics are as yet inconclusive but many writers from differing fields of study, including William Gunderson, research psychologist in San Francisco and Keith Ward, British philosopher, agree with Oxford University mathematician Roger Penrose and quantum physicist David Bohm that the future of science will be focused on integrating consciousness and the manifested material realm. David Bohm, as we have said, believes consciousness and matter are together enfolded within the Implicate order.

Consciousness, and thought produced by the brain, cannot be equated. To accept the concept that the human brain creates reality is to deny the bigger picture. Every thing and every event is a manifestation of the implicate order or super-consciousness, expressing itself at any given moment. This occurs as potential is transformed into actuality in the process of creation itself.

David Hawkins says our perception of events being caused in time is analogous to a traveller watching the landscape unfold before him. But to say that the landscape is being caused by our awareness is merely a figure of speech; nothing is actually unfolding, nothing is actually being caused. There is only the progression of awareness of what already is.[7]

This is what esoteric traditions called "creation by subtraction". The "creation" of the manifest physical realm involves subtraction from the infinite potential which is absolute and formless. As mentioned previously, our brains filter out everything but consensual finite reality or everyday consciousness because the frequencies of the brain do not easily resonate with frequencies of infinite Consciousness. What we call reality is <u>stepped down from universal consciousness into duality</u> in order to allow our brain structure to actually experience it in the physical.

It takes the development of internal mechanisms through meditation, or even occasionally through a trauma, to catapult one through what Joseph Chilton Pearce called "a crack in the cosmic egg" and allow an awareness of that consciousness.

Many years ago my husband, Ian, experienced just such an extraordinary state. One night while studying, a lacewing insect landed near him under his reading light. The lamp highlighted the beautiful pattern of veins in its wings and the segments of its body. As he gazed, he "saw" this miracle of creation, so small, so fragile, so beautiful and so perfect in design. A feeling of love for the creature arose, and then quite suddenly his ordinary consciousness stopped and he experienced a state where he and the lacewing and everything around became one. The illusion of separation had given way resulting in a moment of profound unity consciousness.

At the age of sixteen, Frenchman Stephen Jourdain, highly respected throughout Europe as a mystic and philosopher, experienced a radical awakening while contemplating the famous Descartes statement "I think, therefore I am." As a result of this fundamental change in perception, Jourdain's identity was transformed and he realized in a flash that in truth fundamentally we are pure Consciousness. We are in essence part of creative Consciousness but our brain, by acting as a filter, produces a pale copy of reality so that we live in what Jourdain calls a "state of permanent hallucination."[8] The human brain "creates" our reality only in the sense that it projects interpretation and meaning which provides us with an experience of that part of reality for which we have receptive devices, and does so only in conjunction with a lot more than the brain.

To say that positive intent is responsible for our experience is to severely overstate the power of intention and severely understate the complexity of both science and spiritual tradition. The brain co-creates your reality, but so will brain chemistry, culture, level of consciousness, family dynamics, crises and disasters, environmental toxins, as well as the vicissitudes of everyday life. Causation, then, is a composite of a totality of determinants. All occurrences are multi-determined. "Events", Hawkins says, "are the products of propensity, facilitation, timing, probability, potential, likelihood, alignment, momentum, promulgation, selection, randomness, condition, control, favouritism, public sentiment, the weather, economics, morale, political climate, supply, need, affordability, emotional climate, social morality" and much more.⁹

Social creativity researchers Alfonso Montuori and Ronald Pursuer (1996) describe even more factors that influence our creativity. These include the environment, economic and educational resources, cultural forces, political and organizational contexts, biological factors, psychological and personality issues, and interpersonal relationships.[10] I think he has made his point.

Many factors cooperatively participate in generating the world we observe. We are along for the ride with all of the evolutionary processes. All of creation is an inter-related and inter-dependent whole. Chaos theory declares that in a dynamic system such as ours, a butterfly flapping its wings in Brazil might set off a tornado in Texas.

The bottom line then is, we are not exclusively responsible for the outcome of our intentions or for creating our reality. In reality the truth is far more extraordinary and wonderful. Human experience is part of a larger dynamic of creativity all of which is constantly exerting influence on what manifests on the planet. This process interconnects us with everything within the evolving universe. We are co-participants in a creative process by virtue of the fact that spirit resides within us as it does in all sentient beings. Truly we are spiritual beings having a human experience.

**Our Level of Perception**

We have learnt that we operate on the world but do not fundamentally change its existence. In fact we translate what we see and project our meaning onto it. In this sense we create our own picture of our reality, a virtual reality. Each individual brings to his or her perception, all their own previous experience and so each perceives a slightly different picture of what is observed - or vastly different pictures as in the case with Hitler and Mozart.

Dr. Dean Ornish, president and founder of the non-profit Preventive Medicine Research Institute in Sausalito, California, and Clinical Professor of Medicine at the University of California, San Francisco, describes this as "creating a certain kind of reality". We basically live in dream within which we gradually become aware of more and more of Reality or All That Is, until we enter a state which in meditative traditions is called "Awakening" or becoming enlightened. Your particular dream, he says, is entirely consistent with your own level of awakening which confines your perception and response to that reality.[11]

Brugh Joy, MD and powerful teacher who died in 2011, expressed the same idea, naming these dream states, "state-bound awareness". State-bound consciousness means that the information known to one perspective or state of awareness is bound to that state, in other words, those whose consciousness resonates with that particular state of awareness cannot generally access other states.[12]

David Hawkins has made a study of these state-bound perspectives in books such as "Truth Vs Falsehood", "The Eye of the I", "Power Vs Force" and others. In "Power Vs Force" he expands on the concept of a creative force of unlimited energy and potential out of which manifestation is "actualized" as the phenomenal world and for which he has yet another name, "The Divine Matrix". This Matrix differentiates into various levels which exist as various dimensions expressed in the physical of infinite consciousness.

The implications of Hawkins' decade-long study throw a completely different slant on human freedom of will and behavioural/mental capabilities. David Hawkins divides human consciousness into hierarchical strata, ranging from low frequency states to the highest level of consciousness such as that of the avatar or of enlightenment itself. The human mind is in a continuous state of evolution and development within each substrata or level of consciousness. The brain acts as a radio receiver attuned to various levels within a spectrum of consciousness each resonating at an identifiable range of frequencies. The frequencies within each level create an attractor field [or act as a magnet] influencing the mind-sets, perspectives and even the experience of events and phenomena within that particular level of consciousness.

And it is here, within our level of consciousness that the future of our planet resides.

## 11. THE PERSPECTIVE OF THE PERCEIVER

We were told to park our Mini Minor as close as possible to the car in front. Another car then tucked us in at the rear. Two Aussies whom I had met on the sea voyage from Australia to the UK and myself were crammed into the Mini in a parking area at the Londonderry dock along with hundreds of others awaiting a ferry to take us back to the UK.

It was the time of "the troubles" in Northern Ireland and we three girls on our journey around the British Isles were caught up in the scramble of Irish eager to leave a country in turmoil. We three were to spend two nights cramped in the Mini as all the accommodation had been taken near the ferry terminal.

We knew little of the Northern Ireland unionists loyal to Britain and mostly Protestant or the republican nationalist who wished to remain a separate nation. All we wanted was to escape a city where bombs were exploding around us and gunshots could be heard nearby. The city was in chaos, some of Bogside was on fire. The tension was palpable.

Many years later my husband and I returned to Ireland only to find Belfast has a wall separating Protestant from Catholic areas each with their own schools and churches. While all appeared to be peaceful, the undercurrent of bitterness was still evident in the murals on buildings, a reminder of the bad times decades after "the troubles".

## How Such Hostility Evolves.

Animosity such as the Irish situation begins with the human tendency to characterise people as "us" and "them", "we" and "the other", allies or adversaries. Judgments are made based on the others' behaviour, appearance, race or religious beliefs. These judgments are qualified with distorted and often inaccurate information which validates one's own and invalidates the other's world-view.

Even though the perceivers' views are themselves often one-sided, inflexible and prejudiced they persist in seeing the other as uncompromising, dangerous and even threatening. Assumptions are projected onto the other so that one is incapable of accepting even the more positive intentions of the other. Thus people become separated into opposing camps.

This is just as true of other counties as it is for Ireland. While living in the US it was obvious to us that the first question asked by an American to a stranger is which political party they supported. It was only then the stranger could assess the level to which they could trust the stranger and the amount of cooperation they were prepared to give. Allegiance to a group filters perception so that viewpoints and opinions become misguided and the opposing side is perceived as suspicious even though no individual member might have ever crossed their path.

Responses become stereotyped revealing a sense of loyalty to the viewpoints of the in group. Evidence is selected from sources which provide evidence that confirms their beliefs. Various media channels sanction these strong stances by presenting primarily one side of the issue which may have only a tenuous correlation with the truth. This information is absorbed unexamined by the patrons of that channel or news source. Each different viewpoint has certain strong parameters and opinions are fixed within those confines. So armed with the knowledge that they are "right" even despite evidence to the contrary, they deem further examination to be unnecessary.

This scenario may appear extreme but it applies in some form to opposing political or religious camps in most countries and is consistent over time. It is indicative of the constraints imposed by specific levels of consciousness on the human psyche.

## Hawkins' Levels of Consciousness

David Hawkins proposed that each specific level in human consciousness outlines the content, meaning and values the person experiences. As the levels are progressive, each contains unresolved issues relevant to that field which individuals must resolve before progressing. Each level contains an energy field which defines the limits of awareness and understanding of those within that level. It also organizes the perception as well as the patterns of emotional responses to those events, each particular level having certain behavioural characteristics within it. This explains why it appears impossible to understand the mind of a serial killer or the motivations of a child molester but you also cannot enter or even imagine the advanced mind states of an avatar.

Each consciousness level determines and confines, to varying extents, the perceptual and intellectual positions or viewpoints as well as the psychological makeup and emotionality of those within its limits The energetic restraints of that particular level will govern the way they reject or accept, place values of good and bad, react or repress, and ultimately physically reflect their level of consciousness.

Intentions of those within a particular field draw upon the energies of that field and so determine responses specific to that field's limitations as well as their highest potential. Each choice determines those outcomes as a result of an impersonal mechanism operating automatically on that particular level. Choices then are free only within the boundaries of that level of consciousness.

Paul Bailey has similar approach to consciousness as Hawkins. Bailey is an interesting character who, among other things, has meditated for days in dark caves accompanied by snakes, travelled with gun smugglers across the Sulawesi Sea, witnessed a helicopter attack in Mindanao, been attacked by Tasmanian tigers and very nearly fallen into an active volcano. Born in Australia, he studied spirituality and healing in China, the Philippines and Brazil as well as in Aboriginal territories, and parapsychology in California.

In his book, "Think of an Elephant" he divides human evolution into eight horizons or levels beginning with narcissism and inability to take responsibility for the consequences of their behaviour, to the beginning of a capacity for empathy, self-observation and self-control, advancing to altruism, delayed gratification and certain behavioural insights and then to becoming adult, artistic-creative, psychic, a spiritual elder and leading eventually to transformational super-wakefulness.[1]

As in Hawkins' model, each level presents characteristic stumbling blocks or challenges to the attachments, perceptions, behaviours attitudes and pride of those within it. Hawkins' consciousness levels coincide also with the hierarky in perennial philosophy, and correlate with emotional and intellectual structures in sociology, clinical psychology and traditional spirituality. Hawkins says his studies also correlate well with modern-day scientist Rupert

Sheldrake's morphogenetic Fields hypothesis as well as Karl Program's holographic model of brain-mind function, where in a holographic universe the achievements of every individual contribute the advancement and well-being of the whole. Yet another overlap between science and the perennial and spiritual traditions.

Within our level of consciousness, we become to varying degrees identified with, mentally entranced by and therefore caught up in the content. We usually believe our perceptions to be the truth and we model our way of living on them. It is what Hawkins calls our "positionalities", i.e. our opinions and judgments that lock us into our level of consciousness and block future development. Hawkins says you cannot at the same time hold strong positions and transcend. It is the opposites after all, for and against, better and worse, good and bad, that hold us in the grasp of duality.

Our evolution is therefore dependent on the subordination of our will and our belief systems so as to allow progression through the consciousness levels. By developing equanimity in the face of all that life presents, or in the terms of perennial wisdom, by surrendering and developing a "witness consciousness", our Spiritual Will/intention is free to align with higher dimensional creativity processes. Then, as Lyn McTaggart, would say, it is possible for the receptive neurons in our brains to expand to become receptive to a larger number of wave-lengths in the universal field of awareness she calls the Zero Point Field.

Ian Lawton looks at our place within these levels from a different scientific viewpoint. Lawton is an Australian, educated in London, is the architect of "Rational Spirituality", a movement which proposes that God can only experience material reality by living in and through we humans and all beings everywhere. He finds this esoteric belief has compatibility with String theory in which different configurations of strings effectively produce different harmonic chords that power the strata or levels within creation itself, certain harmonies attuning to various fundamental vibrations within the universe in a type of divine synchrony. [2]

## How Then Do We Change Our Level of Consciousness?

The Nobel Prize winning physicist Max Planck spoke of life-changing windows of opportunity or "choice points" in one's life, where opportunities to raise our level of consciousness occur. Quantum physics, always mind bending, suggests that the opportunity to redefine outcomes may come only at specific intervals where the roads of time bend their courses and approach other roads. [!] Sometimes the roads move so close that they touch one another providing a "choice point". This is like bending the waves of time, according to Gregg Braden.

Braden's conclusions bring the physics into perspective for those of us unaccustomed to mind-bending. "Rather than forcing solutions upon the events of our lives, we are invited to choose which possibility we identify with, and live as if it has already occurred. It is not our will but our willingness to attract the quantum possibility of higher values such as forgiveness, compassion and peace that will attract a higher consciousness level bringing with it a future that reflects such qualities." [3]

The critical thing about raising our consciousness level is this; as our level of consciousness rises, we become progressively more concerned with fairness, balance, protection of the vulnerable, the quality of the environment as well as for the dignity and rights of individuals, all of which are symptomatic of humanity's evolutionary progress. The process of evolution is not contained in the body or the DNA but in the quantum field. It leads us from immaturity to expanded or unity consciousness. There are clearly areas of commonality between mystical experiences of unity-consciousness and what physicists describe as the universal field of awareness. It is this evolution, innate to the overall Field of consciousness, which Hawkins believes guarantees the salvation of mankind, and with it, all life.

On a global scale, cultures, civilizations and communities develop a collective code of accepted behaviours just as individuals do. Each has its own level of consciousness determining the way they see the world, think and behave, whether they treat their population humanely, their attitudes toward war or peace, and whether they demonise a minority within their community. Even minorities within cultures contain contrasting levels of consciousness existing side by side with mainstream beliefs and behaviours.

Civilizations are peaceful, prosperous, create war or become poverty stricken depending on the dominant level of consciousness at the time. Law-abiding societies contain pockets of corruption and white-collar crime; some sections of otherwise devout religious societies favour terrorism; the Amish, seek a wholesome lifestyle by isolating themselves from the technology of the modern world that surrounds them.

Hawkins' map of consciousness begins with world-views centred on evil, blame and humiliation, [calibrated at 20] and climbs to enlightenment measuring 700-1000 and characterised by Beingness. Humankind has at last and only recently reached the level of integrity or courage at 200 where a critical number of people put back into the world as much energy as they take. Now that the level of consciousness of mankind has passed the 200's, collective society is becoming far less self-absorbed, war-mongering and focussed solely of gain. It is developing progressively more humanitarian, caring of others in need and appreciative of species other than ourselves. This gives us some hope that the Tipping Point will become our Turning Point.

However, we cannot become complacent. While humanity is evolving, we nevertheless remain in a vulnerable position. There are no guarantees that uncivilized minds within certain cultures will not act to send society backwards in spiritual time to a darker age. The fact that humankind as a collective has reached a level of integrity, means in truth that a small number of people at higher levels are balancing the majority at much lower levels; most of humanity is still caught up by lower states. Having passed the 150 level with its hate, antagonism and anger, our consciousness still must reach 400 before our life-view becomes meaningful and we achieve reason and an understanding of spiritual values.

We began this chapter with the description of our perception and the way it dictates our worldview. We now see that it is our level of consciousness that is the attractor force drawing to us events and experiences that resonate with the world-view of that particular level. In order to progress as a species the emphasis for each of us must move toward non-judgmental attitudes, contributing to the happiness of others, away from getting and having and toward giving and evolving. Our evolution takes us from ego-identification and "us and them" mind-sets toward cooperation with one another and affinity with the planet as a whole.

The future rests with us, our awareness, our conscious choices and the world-view we carry. The collective level of consciousness is raised by changing our individual belief systems and making personal choices with integrity. By refusing to identify with judgments and positionalities and choosing a higher quality of thought, feeling and emotion, we attract a higher energy field into the collective consciousness. Thus we as individuals set a new course and a new outcome for humanity as a whole.

The new theoretical physics informs us that everything in the universe is connected to everything else. Our decisions then, have consequences far wider than ourselves. Our choices reinforce the formation of powerful Morphic-fields, which are the attractor patterns that automatically influence others within them.

Every life-supporting decision we make supports all life within the whole interconnected ubiquitous energy field. Our own responsibility as participants in life correspondingly becomes of vital significance to the whole of life. Elevated human consciousness and our loving thoughts and compassionate heart state are all therefore fundamental to eliciting the synergy of the higher frequencies of the universe that in turn fuel the evolution of the planet, its inhabitants and the environment that supports us.

We affect our entire universe, not with our egoic intentions but with the love or hatred, the compassion or grievance we hold in our hearts. Ultimately we change the world not simply by what we do or say, nor just by our thoughts or intentions, but as a consequence of the quality of the consciousness we bring, moment by moment, to every aspect and event in our lives.

This is the message we must take from the crises that are threatening our planet at this time in our history. It is the moral lesson we must convey to the next generation. For this reason the quality of our consciousness and the love or animosity we bring to the next generation is crucially important. We will investigate the power of love in the next chapter.

## 12. LOVE AND THE HEART

In my youth my mother was a kind, dutiful and disciplined parent, nowhere near as austere as my grandmother who reminded me of Queen Victoria in both appearance and manner, but certainly far from the new age mother who attends to every emotional need of her child. My brother and I were to be "seen and not heard" and told to go out to play as soon as a meal was finished. There was no thought of entertaining children or becoming a chauffeur, a teacher or playmate as is common today.

At 13, I remember saying "No" to my mother for the first time. Thinking back, I could have chosen a more appropriate moment than the presence of my father to launch my attempt at independent thinking. My father exploded into action as if to an enemy invasion. He mobilised into a pursuit that conveyed us several times around the dining table before fear overcame me and I succumbed to a severe beating and was thrust into my room to consider my sins.

Some months later it was with absolute shock that I realised my cousin Ken had suddenly burst from childhood into young-manhood. To my amazement I noticed that his relationship with his mother had changed completely. Ken was now at least a head taller than my aunt and he was actually putting his arm around her and fondly teasing her, something she obviously enjoyed from her only son. The impact of this occurrence has remained with me until this day. It had never occurred to me previously that children could "be friends" with their mothers, or especially, that they could have a sweet, teasing relationship with a parent.

I decided from that moment that I wanted to relate differently to my mother but could not imagine how that relationship could be. I certainly could not put my arm around her. Mum did give me a kiss goodnight but at that time we never otherwise touched or cuddled.

I began my campaign by asking her if I could read my Geography book to her to help me learn for an exam. Mum had had very little education because she was a girl, one of many in her family and her husband, my father, a product of his time, did not approve of her working for a living so her horizons were fairly limited. To my childish surprise, as I read Mum began to take a lively interest in the subject matter. This was a bonus that was completely unexpected. Mum asked me questions and I felt very grown up explaining various landforms and weather patterns to her. For one assignment I remember studying penguin behaviour and we laughed together at their antics. I later asked her about wildflowers for another assignment. Before long we were enjoying our afternoons together.

When I needed clothes, instead of mum buying them while I was at school, we would go together to the local shopping centre, - a much-anticipated outing. We even began to go to "town" [Sydney] together and have a cup of tea at Repins like real ladies. I had reached out for a new bond with my mother, no longer as an obedient child with her parent, but as "friends". From that time until she died, we so enjoyed one another's company that each time I had to leave to go to my own home, we both shed tears.

It would be years after my teenage years before my relationship with my father, a philosophical man, nature lover and a great reader of classics, began to change as well. My father had been humourously scathing of women driving a car, reading the news on TV, going to work, even wearing trousers. As an impressionable girl I assumed that he felt I was inferior. It took my university attendance and my various adventures overseas to turn our relationship around. I felt determined to prove to him that I was OK in spite of being born female, and as I did, our relationship too began to change or perhaps I may have just began to feel more acceptable to him and felt closer to him as a result. Relationship is hard work – it takes time and an open and determined heart along with a bit of personal soul searching.

My relationship with my mother gave us a very strong heart bond. Many times Mum and I would try to ring one another simultaneously. Several times we bought uncannily similar gifts for one another at Christmas time, and we had a "knowing" about how the other was feeling even when separated by a distance of thousands of miles. It was a bond that united our hearts until the day she died in my arms. After her death, I felt as if a part of me as physical as an arm or a leg had been severed.

Not only have molecular biologists discovered that the heart is the most important endocrine gland in the body but neuro-cardiologists have found that, rather than being made up entirely of the muscle cells essential for moving blood throughout the body, 60-65% of heart cells are actually neural cells! These cells are identical to the neural cells in the brain, operating through the same connecting links called ganglia, with the same axonal and dendritic connections as well as through the very same kinds of neurotransmitters found in the brain.

Researchers at the Institute of Heartmath in California have found that the heart is the most powerful generator of electromagnetic energy in the human body, producing the largest rhythmic electromagnetic field of any of the body's organs. The heart's electrical field measured with an electrocardiogram [ECG] is about 60 times greater in amplitude than the electrical activity generated by the brain. In fact biophysicists have discovered that the heart is a powerful electromagnetic generator that profoundly affects the brain. In an interview in 1999, Joseph Chilton Pearce said that the heart furnishes the whole wave spectrum from which the brain draws its material to create our internal experience of the world.[1]

As we have found, our biology is not controlled solely by molecules, atoms and biochemistry. The human body is part of the interconnected network of information and energy that makes up our universe. Australian researchers, Peter Fraser and Harry Massey have found that, although genetics and cellular chemistry are important facets in the way the body functions, "there is a deeper reality to the body, one in which physics, especially the field of quantum electrodynamics, governs physiology. The interaction of quantum waves imparts energy and information that is encoded in the human body-field, which serves as a holographic template for the physical body." [2]

In the world of Quantum theory, all particles are constantly interacting with the myriad background energy fields making up the Zero Point Field. This is the "non-local substrate" [universal energy field] that feeds the body, as it does the entire physical world. The Zero Point Field creates a medium or underlying mechanism enabling the body's molecules to speak to each other non-locally and virtually instantaneously. Channels or energy meridians transfer the information which drives the biochemistry, a fact well-known in traditional Chinese medicine. Here again ancient methodologies meet modern physics.

We think of the heart as a pump driving blood throughout the body which is, of course, a vital service to the body. But scientists have discovered that the heart has an even more miraculous function. The heart actually uses frequencies of all varieties as an information device in a whole-body feedback and communication system. This frequency system operates through waves of energy which encode and transfer the information that drives the whole body's biochemistry.

As we have found, German biophysicist Popp in 1974 found that DNA stored energy in the form of light. Particular frequencies coherent with the heart's biophoton [light] emissions resonate with corresponding molecules in various organs, imparting information to those cells. Then something truly miraculous happens. Each individual cell becomes energetically aware of its unique functions and cooperates with all others of a similar type in order to become and to operate as a specialized organ within the body! In the same way the single cell of an embryo multiplies and becomes differentiated into each of the body's systems, beginning with the heart and eventually becoming a complete human being. This must be the greatest miracle of all the wonders in human biology.[3]

The early development of the heart is essential because the direction and organization of frequencies from the universal field within the body is carried out by the heart. According to scientists at Heartmath in California, the heart's energy field is shaped like a torus or doughnut. This torus lives within a nested hierarchy of toroidal systems. The significance of this is that the torus function is holographic so the heart is able to mediate between our individual self and universal fields in a holographic way. When conditions are right, the heart's electromagnetic field can resonate at the same frequency as the unified fields of potential or Zero Point Field [ZPF]. In this way the heart draws on universal intelligence and conducts it for application by the body's systems.

As if this is not miraculous enough, the heart is apparently also responsible for the moments of insight-intelligence as experienced by Mozart, Einstein, Laski and others. The heart's intelligence is not verbal or linear as is the intellect in our brain. It is holistic and manifests in a gestalt or complete knowing.

Toronto paediatrician, Dr. Bernard Laski, agrees with Pearce that in order to access insight intelligence, the mind must be clear and the neural circuitry ready. This requires the circuitry of the right hemisphere, connected as it is to the heart.[4] The brain merely interprets the whole answer linearly, so the information can then be translated from the field of potential [ZPF] to the brain.

It is when we are less centred in the cerebral hemispheres and more conscious of the subtle energies of the heart through meditation and compassionate action that we become capable of attunement with the universal life force. In so doing we allow the heart's wider source of wisdom to instruct the earth-bound brain. Heart energy, connected to ZPF synchronises all our bodily systems within the universal hologram and even enables superhuman feats to occur; telepathy, clairvoyance, intuition, guidance and precognition.

The body's instrument for receiving the information-carrying waves from the larger non-linear field is, as we have said, primarily the heart, because its frequencies are more "in synch." with the higher frequencies than the brain. The information-carrying frequencies can be conveyed throughout the body because the heart and limbic systems have neural connections with the right hemisphere and prefrontal lobes.

Findings from heart donor patients confirm that this transfer of information is carried within the body's energy field. It can be carried from heart donor to recipient within the heart's energy field so that the receivers of the new heart report different bodily information typical of the donor; different feelings, thought patterns, cravings and even personality characteristics have frequently been reported.

Paul Pearsall, clinical and educational psychologist, commenting on this energy received by the heart, says, "What we call mind, consciousness, or our intentions are really manifestations of information-containing energy. What I am calling "L" [love or life] energy is the basic code of life and what our "system" remembers as 'who' we are. The heart is the primary generator of info-energy. Because we are manifestations of the info-energy coming to, flowing within, and constantly being sent out from our total cellular system, who and how we are is a physical representation of a recovered set of cellular memories." He adds, "Energy cardiology suggests that the heart is the conductor that keeps all the cells playing the same score." [5]

This means that the blueprint of us as individual beings, is carried in the universal or Zero Point Field, received by the heart and conveyed to all the trillions of cells throughout the body, resonating individually with the particular frequency of each cell. "L" energy, riding along within this universal energy field and being sent throughout the body, is then the ultimate syntropy upon which our wellbeing and ultimately our evolution rests.

**And its Energy is Love.**

We have always associated the heart with love and it is through love that the connection with the universal energy field is created. Professor William Tiller, Professor Emeritus of Stanford University's Department of Materials Science, writes, "Elevated and loving human consciousness is a likely key requirement...we create our collective future via our thoughts, attitudes and actions [by] maintaining an uplifted spiritual/mental/heart-state. It is the practitioner's love, compassion, devotion to service and intent that can elicit the 'unseen' assistance of the universe." [6]

Through this connection we begin to manifest what you have become rather than what we think. While the mind's thinking process creates good and bad because polarity is its nature, the heart has a direct connection with universal energy or God-force.[7] Sara Paddison of the Heartmath Institute in Boulder Creek, California, writes "Whatever your religion or cherished beliefs, the heart is the access point in the human system for experiencing God." Both Western and Eastern traditions teach us that Love is the energy of creation.

Love energy has been confined in our minds to the warm and fuzzy. **In fact love is the essence of the universal energy field itself; a non-local, creative, organizing and all pervasive energy that counters entropy as it sustains, connects and integrates all other energies.** This life force centred on the heart is the basic or primary force behind all systems. Paul Pearsall, who received numerous awards for his research on the relationship between the brain, heart, and immune system, believes that "the subtle energy of the heart and the cellular memories this energy creates, are the missing pieces of the most sacred puzzle – what is life and what is it for? Tapping into this energy, he says, answers those crucial questions." [8]

Dean Ornish, mentioned previously, agrees. Love, he says, is the fundamental attractive power. Newton saw gravity as a force which interconnected the universe but it was the expression of the universal love of God that formed the glue that held it together. The love carried in this God-force or field, he says, exists in all systems from the microcosm to the macrocosm and brings it all into balance.[9]

The Universal energy is filtered down and diluted by our normally undeveloped brain and emotional systems to enable it to be available to our neural system. A heart, when closed off from the larger universal energies, sends limited and more earthbound frequencies to the brain turning love into fear. The whole system loses its equilibrium and well-being. In doing so it cuts us off from the larger life force. In contrast, an open heart allows the energy to flow and nourish our system. It allows syntropy to occur as intended.

We are like a radio and our bandwidth can expand or contract. Under certain conditions, Lazlo says, the receptive patches in our brains become more receptive to a larger number of wave-lengths in the ZPF. Other external conditions also limit our receptivity. Popp, for instance, examined the effect of stress on photon conductivity. In a stressed state the rate of biophoton emissions went up and equilibrium and coherence was affected. When people establish a relationship, their brain patterns become either highly synchronized or incoherent and the effects of both contrasting emotional bandwidths are carried instantaneously to the other.

### The Heart and Child Development

It is not coincidental that the first organ to form in the fetus is the heart. It develops first in order to provide the electromagnetic spectrum that instructs the DNA. It is no longer believed that the genes in our DNA are set into permanent or fixed programs. In fact they too, as mentioned earlier, are greatly affected by the environment in which a child grows, especially the emotional environment, for which the heart is the receptor organ. Emotions can actually change the heart's electromagnetic spectrum and influence the DNA as the child grows and affect the brain's ability to learn and thrive. The emotional atmosphere that the parents create is vital to the child's wellbeing.

Peter Fraser and Harry Massey, Australian scientists and authors of "Decoding the Human Body-Field: The New Science of Information as Medicine", stress this point. As an embryo develops, they say, the organs create what they call "Energetic Driver fields" which impart basic constitutional energy and information to the embryo's body-field, and hence to the physical body.

The infant's body-field is aligned with fields of the parents, especially the parental emotional environment, which affects all aspects of the bio-energetic health of the child's organ systems and the body in general. All bodily structures are therefore influenced by the emotional environment, especially for the growing child, and form the basis of the youngster's mental physical, mental and emotional health throughout its life.

The ability of the parents to attune with higher frequencies then is critical in the complete development of the child. Our future then rests on the degree to which brainwaves are synchronised or incoherent with Universal energy as they are carried between parent and child beneath our normal awareness.

## 13. THE CRITICAL ROLE OF PARENTING

After leaving my first marriage I took my $20.000 settlement and moved to the country to live the "back to the earth" existence I craved. With resources limited to such a small amount, I decided that owner-building was the only way I could afford a home. In spite of the attitude of building contractors in my country area, aghast at the thought of working with an unmarried woman who was presumptuously owner-building without the guidance of a man, I summoned every bit of my newly-won assertiveness, added to it my counselling communication skills, and strode valiantly into an occupation I knew nothing about.

With long commutes to my extensive school-counselling district, I had to stretch my energy for the two terms necessary to complete the construction. I then breathed a huge sigh of relief. I moved into my beautiful house and settled back to enjoy the peaceful country existence I planned.

It was at that exact point my future husband appeared, bringing with him two young boys traumatised by the breakup of their parents not long before.

I had wanted children for years, so was eagerly awaiting the joy of having two young ones delivered so effortlessly into my life. I set about preparing rooms for them and investigating outings to make their lives enjoyable while with us and discovering what young boys might like to eat.

The boys, on the other hand, then six and three, were faced with a new unsettling set of circumstances on top of their previous distress. Their lives met with sudden upheaval yet again and they were not at all delighted by the idea of weekends spent with some interloper instead of their mother.

Having very naively expected my eager excitement to be met with something approaching enthusiasm, I was stunned by the stony faces and cold reception I received as they arrived at my back door. When eventually they could build the courage to speak to me, the frank and forthright expression of young boys, too, was unforeseen. I began to stagger under their silent resentment of the stranger who was apparently the cause of their distress.

"You are not my mother!" "I don't have to do what you say!" were the biting words which I only later saw as completely normal, fearful reactions to new and frightening circumstances thrust upon a young life. They had been removed from their secure comfort-zone and had met an adult similarly insecure and fearful about the unfamiliar situation being thrust upon her.

I felt stunned and alarmed by the way my naïve anticipation had turned and struck me in the face. I alternated between tears and feeling unjustifiably hard done by.

Sometime later I caught a glimpse of their little faces while they were off-guard and recognised their fear and confusion. My heart went out to them. It was easy then to empathise with their distress and to promise that although I would never replace their mother, I would try to be a secondary mother figure who would love and care for them to the best of my ability.

As I got to know them it was not hard to give my love to them. I now feel blessed to have them in my life. Love has healed all the fear. Now in their adulthood, I am very grateful that they entered my life even by "the back door" as they did.

It is critical that parents-to-be expand their understanding, tolerance and love because these critical factors influence the health, wellbeing and even the future possibilities of the developing child.

It is the emotional state of a mother-to-be that profoundly affects the fundamental heart energies, which in turn affects the foetus' DNA and consequently the brain and the whole of her infant's development. Consequently, when the infant is born, it must have emotional stability, compassion, nurturing, benevolence and caring as a foundation for the child's intelligence, growth and development, memory, health and well-being. With these factors in place, the child's frontal lobes will develop as designed, allowing the possibility of connecting with the heart energy, and so ultimately contribute the next step in human evolution.

It is well known that heavier babies are born to anxious fearful and stressed mothers-to-be. Stress affects the energetic structures of the unborn child who is born not only with heavier musculature but enlarged hindbrains, the part of the brain that deals with flight or fight and reduced forebrains, the part capable of higher functioning.[1]

Some western birthing methods add to the stress of the infant who may then be packed off to day-care where she suffers further abandonment at an age where stability of caring is crucial. Threats, punishment and prohibitions of many kinds all reinforce the more primitive reactive hindbrain's dominance so that the child does not have full access to the higher intelligence of the forebrain.

The mother's emotional state will affect the neural growth of the developing child which can shift between a defensive, combative stance to one that is reflective and insightful accordingly. Dean Ornish says, "If we don't get enough of the normal nurturing as children our brain serotonin systems don't develop normally. We grow up to have a brain that is more likely to be sensitive to forces that cause it to be depressed, hostile and socially isolated."2   He describes a number of biological mechanisms through which a lack of love and happiness and an excess of hostility and anger affect the child's biological processes that can lead to disease and even premature death.

No pressure then. Parents just need a large dose of super-human resilience, tolerance, understanding, forgiveness, not to mention love along with the consciousness that their mental state is critical to the optimum development of the child.  And of course these are the very qualities the human race need to wean itself away from; intolerance, bigotry, blame and lack of compassion. It is the choice of reaction that we ourselves bring to any situation that to a large degree determines the emotional development of the child.  We become more compassionate by using the "opportunities" that our children habitually present to us to practise our own evolution as human beings.

Children have an ability to ignite frustration, irritation and anger just by being children. It's part of their job description. Many of our loved ones can have the same effect. It's what they're there for, to help us learn better ways of reacting and being.  Living intimately with people gives us a unique opportunity to see the consequences of our actions and our words. Words are extraordinarily powerful and our words can positively or negatively affect the development of our child's mental and emotional wellbeing.

Children are impressionable and they take their clues from us.  It is our ability to tune into what is really going on beneath their behaviour and our understanding of their struggles to be OK and lovable as well as a good dose of generosity of spirit rather than our knee jerk frustration that will encourage self-confidence, a willingness to cooperate and to please.  Our habitual criticism and put downs on the other hand, have the power to cause timidity, lack of confidence, a sense that they can never please, a lack of initiative, a fear of making a mistake and even a feeling that they are hopeless and helpless. Frustration and anger ricochet back on themselves as the child picks up the feelings and reacts to the stress with a defensive response. It is the encouragement and nurturing of a child's best qualities that will support and foster their best qualities and assist their maturation into the best person they can be. Our job in any relationship is to develop ourselves so that we become the person upon whom our loved ones most wish to model their lives.

John Bradshaw, American educator, counsellor, motivational speaker and author, says, "What is clear is that we must accept that caring for children in their earliest stages of development is the most vital concern any society has to address. The major source of violence in any culture is the abuse of children." And abuse is carried out by those who themselves have been damaged.

Much of the violence in the world today is rooted in inadequate parenting by overworked, mentally fatigued, emotionally unstable and often substance affected parents with non-existent parenting skills.3 Montessori, the great educator believes, and it is my conviction after years of counselling in schools, that it is the personality of the parent, his or her need to control, the unresolved issues that influence preconceptions as well as faulty unexamined beliefs that makes for the most detrimental element in the child's environment.

One of the greatest gifts we can give our children is our own growth so that we can better influence our children's moral intelligence and wellbeing. We need to do the work of discovering our true self by rooting out our unbalanced natures, unresolved issues and unexamined beliefs. Children need parents who are authentically present and balanced because whenever the child's heart perceives negativity, hostility or danger, it drops out of its harmonious mode into an incoherent one, triggering the release of the single most damaging hormone, cortisol, the stress hormone.

The number of damaged children, underachieving children, discipline problems, anorexia nervosa and even suicide in young people is rapidly increasing. Our children have been signalling us for years that something is critically wrong with our child rearing.

Bruce Lipton, the developmental biologist mentioned earlier, believes newborns can experience almost all the emotions of adults and can express rage, jealousy, anger, love and sadness, but it is the parents that actually tailor and shape the child's physiology and behaviour. Lipton agrees with Pearce when he says it is through bonding and in all their interactions that the parents are programming the responses of the child. An unbonded child will almost certainly have developmental problems such as attachment disorder or attention deficit hyperactivity disorder.[4]

This is all sounds like a very big ask for parents. It comes down to two essential criteria; that we endeavour to resolve our own issues that stand in the way of unconditional love, and that we create a calm space with no anxiety or agitation in which the child can feel secure. Stress and anguish are translated into physical ailments, as well as mental and emotional instability. It is a calm atmosphere and tranquility that brings balance and health to the body.

A child brought up in a troubled or dysfunctional atmosphere craves intense emotion and will often seek negative attention from his parents because negative reactions carry with them more intensity. If the child is chastised and blamed, he will continue with the same behaviour in order to draw negative attention from the parent and gain the intensity his psyche has become attached to. It is behavioural limits, applied consistently and calmly, that overcome this trend.

In today's world many children spend less time with their parents than with various forms of technology. Studies have proven video games have a negative effect on the child's brain chemistry. The brain's pleasure reward centres become so over-stimulated that they become desensitized and crave higher and higher thresholds of stimulation. In adulthood this can lead to cravings for drugs, alcohol, pornography, gambling and all other excesses. Children also become desensitized to manipulation and violence and consider them normal behaviour.

It is when children have stimulating family life, discussions around the table, walks in nature, family picnics and outings, participation in healthy exercise and social activities that they grow into balanced youngsters who respect people from all races and beliefs as well as the larger environment.

Redford Williams MD, professor and director of the Behavioural Medicine Research Centre at Duke University Medical Centre, describes a number of biological mechanisms through which a lack of love and an excess of hostility affect the biological processes and could lead to disease and even premature death. "If we don't get enough of the normal nurturing as children our brain serotonin systems don't develop normally. We grow up to have a brain that is more likely to be sensitive to forces that cause it to be depressed, hostile and socially isolated" [5]

Attachment theory, first conceived by John Bowlby, suggests that infants' and children's relationships with their parents develop "internal working models" that colour their expectations of relationships and their overall world view for the rest of their lives. These models shape their belief in whether they deserve care and love and these expectations become filters through which they construct new relationships consistent with these beliefs throughout the whole of their lives.

Children actively seek to imitate those they admire and view as strong-willed. The aim of parents should be to stay warm but immovable, calm but refusing to be manipulated by poor behaviour. This creates a secure environment for the child. Internalised beliefs, based on family history, predict future behaviour; children raised in an atmosphere of warmth, security, love and discipline [rather than punishment] evoke from others responses that conform to these beliefs and expectations. If children are brought up with rejection or fear they will become defensive and invite rejection.

Children look for models who will define their behaviours, attitudes and beliefs for them. Instead many parents in modern times often model a preoccupation with accumulating wealth and an overarching desire to get ahead at all cost. These are the parents who have no time for touching or listening, show little interest in the child as an individual and often have little self-discipline or self-restraint themselves.

The future of the human race is in our hands. Love is the glue that holds the universal energies and it is parents who are responsible for the ability of children of the future to connect with loving energies. The emotional atmosphere that the parents create is therefore vital to the child's wellbeing and ultimately to the wellbeing of society as a whole.

Jeremy Griffith, Australian biologist and author on the subject of the human condition, says it is rare to find individuals who were so adequately loved, nurtured and sheltered from corrupt reality during their upbringing that their "instinctive self" or soul escaped being hurt, damaged or corrupted.[6] To escape being damaged requires a special kind of parenting.

When parents establish a deep loving connection together, their brain patterns become highly synchronized and are carried instantaneously to the child. With these factors in place, the child's frontal lobes will develop as designed allowing the possibility of connecting with the heart energy that will enable the next step in human evolution, away from the reactivity of the hindbrain, with its 'flight or fight" mentality and toward the cooperation and compassion of the Mind/brain.

Parents automatically model the qualities they have acquired and show by their behaviour <u>who</u> they have become as people. "Who they have become" refers to their level of consciousness, rather than what they think or believe. When parents are connected with the larger life force, their heart's inter-dimensional field resonates at the same frequency with the unified fields of potential or ZPF. When the heart and brain are synchronized in this way, we become coherent and open so that the whole system resonates with the energy of the interconnected universe. Once the unified field is accessible, self-control and wisdom can replace the parental control and punishment of the past. What a gift to bestow on our children!

We have the capacity to create a new generation of young people who, raised in optimal conditions, less centred in their hindbrain are more able to attune to the "higher values" by tapping into elevated human consciousness. It is our thoughts attitudes and actions attuned to the frequencies of the universe that are the key to our evolution.

The parents of this world have a responsibility for creating a world of love, all-encompassing compassion, balance and healing for the next generation. Only then will the spiritual and physical progress of humanity and the future of our planet be assured.

## 14. WHAT THE WORLD NEEDS NOW

After leaving my first marriage, I decided to fulfil a dream and travel to South America. I joined a small group of Australian tourists with a Spanish speaking leader and a plane load of locals on an ordinary domestic flight over the Andes from Las Paz in Bolivia to Santiago in Chile. Lunch was finally served and as we were ravenous, we set upon it with gusto.

After an incomprehensible announcement in Spanish over the intercom, the air hostesses began snatching our plates and hurriedly returning them to the hostess quarters. Annoyed and puzzled, we complained bitterly. Then it all became clear. Our leader was able to translate the message which had informed passengers that flight control had been tipped off that there was a bomb on board. As a plane on a similar voyage had been destroyed by a bomb only the previous day, the warning had been taken very seriously by the pilots.

As we neared Santiago, we were told to buckle our seatbelts and that we were being diverted South to make an emergency landing at Concepción. I guessed the possible reason for this was that our plane's crashing there would have less impact than at Santiago, the more populated capital. The message went on to say. . . . we should hold on to our handbags or briefcases so that if the plane crashed, the bodies could be identified! Consideration for the emergency workers is a virtue, I guess, in a situation such as this.

The plane began to circle and ditch fuel. My mind went into overdrive as I anticipated a very loud explosion that at any moment would almost certainly end my life. It seems regrettable I cannot report that some deep philosophical insight or lofty discovery came to me at this time, something worthy of the gravity of my situation. But alas that was not the case. Instead, two things occupied my thoughts to the exclusion of all else. One all-encompassing idea was that if I had to die, ending my life over the Andes seemed more acceptable than fading away slowly or being hit by a bus in my own neighbourhood. The second was acute disappointment that my loved ones would not receive the gifts I had so painstakingly selected for them. The mind is an extraordinary thing.

A tiny terminal became visible and, as no resounding crash had so far occurred, I felt certain that the moment the plane hit the ground, the bomb would be triggered. An ambulance, a fire truck and a jeep containing sniffer-dogs were racing beside us. As I looked down, the ground was approaching at a very fast pace taking me second by second to my death.

Finally the plane touched down and . . . nothing happened! Avoiding the terminal area, the plane taxied quickly to the centre of the tarmac. It pulled to a halt. Still no huge bang. The silence was palpable. There had been no explosion. A voice told us in Spanish to leave the plane as quickly as possible by the rear exit and to take no cabin luggage. As if we needed to be told to move quickly! Only one bus was available. The rest of us ran faster than we had ever run before, across the tarmac to the terminal.

We remained at the terminal for six hours during which time we were asked to go two by two into the plane and remove our cabin luggage, and sometime later when the hold luggage had been removed onto the tarmac and inspected by sniffer-dogs, to go in groups of four to inspect our own suitcases for any unusual inclusions, ostensibly bomb-shaped. It had been a late "inclusion" in someone's suitcase that had blown the previous day's plane out of the sky.

My suitcase fasteners had been broken and so were the fasteners on the suitcase of an Alice Springs hospital matron who was with us. Our cases were separated from the others in an isolated spot. Both she and I were asked to feel into our bags, tampered with as they were, to see if anything foreign and presumably round and hard had been placed there. Having rummaged extremely gently through our belongings well away from danger to other passengers and finding nothing uncomfortably special, we were sent back to the terminal. The airport bar was now supplying free brandy – a nice touch in a situation such as this. Then, suitably numbed, we re-boarded the same plane and returned North to Santiago.

We arrived very hungry and a little light-headed at 11.30 that night. The staff at the hotel had been alerted to our plight and had kindly stayed back to serve dinner for us at that late hour. I remember the food tasting extraordinarily good, the atmosphere particularly welcoming and warm –and our mood jubilant. I recall very nostalgically the beauty of the piano music played with such feeling by a young man who had never been taught to play but obviously "felt" music in his soul. His version of "Nights in White Satin" [Moody Blues] remains with me to this day.

It has often been reported by those who come face to face with death, that life afterward becomes heightened, the colours more pronounced, the music sweeter and the friendships more special. After an experience that takes people to the edge of existence, it is common for survivors to reassess their priorities in life; the insignificant is put aside and what is real prioritised. The little games and petty grievances of the past stand revealed as trivialities and meaning and purpose become the benchmarks for life. Nature, previously viewed as mere background scenery, becomes a living reality, awesome in its seasonal display. Loved ones become cherished as never before. Life itself, so nearly removed, takes on a sacred character. A close encounter with mortality resembles a Vision Quest from which we emerge with a new perspective on life.

I look back on this life experience as an initiation, just as so many other deeply traumatic or even ecstatic moments are in peoples' lives. Such moments are like a rebirth when we leave behind much of what was comfortable, familiar and accepted and begin to see a different vision of our aims and goals and the way to live them as a day by day reality.

This experience ignited in me a profound new sense of being truly alive, of gratitude for life and an inner sense of the blessings I had been given. There emerged an increased awareness of the sanctity of life itself and of keeping every moment precious. It was as if the simple things of life soared in importance, my loved ones, my home and my friends, the freedom from war and tragedy in my life. It would have seemed ludicrous at the time to value highly any sense of social position, status symbols or the state of my bank account.

Trauma, life-threatening situations or illness are often the launch pad for what Richard Moss calls a "radical awakening", when a person perceives in a flash the falsity of their "normal" state of consciousness. It is as if we awaken, at least for a time, from a dream-state which formerly held us in its spell. The psyche becomes rewired so as to instil a reverence for life and for existence as a whole, and a new respect is born for all sentient creatures co-habiting the earth, each of which lives precariously, just as we do.

## The World in Crisis

Our current world situation heralds the need for such a radical awakening, an expanded world view on a planetary scale. Mind-sets, though, are so entrenched into human psyches that many people would prefer to reject the validity of scientific data rather than go through the difficult process of changing ingrained mental patterns. The ego refuses to surrender the grievances and righteous indignation or the cherished positionalities that feed it.

It is just these hardwired attitudes Jack Kornfield, American psychotherapist and Buddhist teacher, addressed in his book, "A Path with Heart". "We must remember that the world's current problems are fundamentally a spiritual crisis," he said, "created by the limited vision of human beings - a loss of a sense of connection to one another, a loss of community, and most deeply a loss of connection to our spiritual values. . . . The worst problems on this earth - warfare, overt ecological destruction, and so forth - are created from greed, hatred, prejudice, delusion, and fear in the human mind." [1]

Maturity does not occur automatically with the reaching of a certain age. It is the acquisition of the "higher qualities" Darwin spoke of as essential to human evolution that raise us from what Einstein called "nuclear giants but ethical infants". It is the values we tend to understand only after a life-threatening or traumatic experience that can break through our delusional beliefs and take us to the next step in our growth toward compassion, kindness and understanding; only these can transcend the inbred greed, hatred and power-hunger that lurk within us and bring transformation to the world.

James Lovelock published a book in 2006 the title of which, I would have thought, would be enough to shock a person into adopting higher values with some urgency. "The Revenge of Gaia, Why the Earth is Fighting Back – and How We Can Still Save Humanity." In it he quite unequivocally speaks of the global change of heart and mind that is needed at this time.

Lovelock gives the example of astronauts like Edgar Mitchell and Buz Aldrin who looked back at the Earth from space and saw not only that it was a stunningly beautiful planet but simultaneously realised our Earth is not something inanimate but a living thing. This realization stimulated a perceptual change that altered the way both these men chose to live from that time. Lovelock believes it is only with a new perceptive such as this that we can instinctively sense what appeared to be hidden previously; that we have made an enemy of our planet and that we must change our hearts, our ways and our mind-sets before the planet begins to retaliate.[2]

**The Beliefs We Need to Change**

It could be said that the world is a stage upon which our egos play out their undeveloped parts. It is the villain in any Shakespearian drama who shows his or her evil and unevolved side traits such as jealousy, hatred, vengeance, vindictiveness or greed, thereby causing havoc and death to other players as well as sorrow to their families – and turning the play into a tragedy in the process. The characters with the defective, culpable and irresponsible roles must develop into to more mature, evolved and noble characters in order to bring the drama to a positive conclusion or die ignominiously in the end.

It is our ego that delights in the role of victor, the manipulator of others or the one who controls and uses power in order to satisfy its undeveloped side. It is our ego too that can play the opposite role of the martyr or the victim, milking every bit of suffering out of being wronged, unappreciated or misunderstood and taking pleasure out of righteous indignation and blame of others. The unevolved and adolescent side of our ego, the narcissistic, entitled side, takes no responsibility for the effects of its actions and displays its defects in every action and word as it wends its destructive way through life.

The ego clings to what David Hawkins calls "positionalities", as if our own viewpoint was equivalent to our identity and giving up our strongly held opinions would mean annihilation to our very being. Resistance, part of the teenage job-description, can too easily become a way of life when it becomes ingrained into our psyches and its natural consequence, contention, is brought into every interaction.

Indignation and long-held grievances serve the undeveloped ego equally well. It is as if some people thrive on indignation in order to satisfy some inner need for ego-building. Some elderly people have resistance indelibly etched into their faces after a lifetime of "sticking to their guns". Refusal to alter one's ways even in the face of impending disaster extends through every level of human interaction, family, community, state and nation-wide and is exemplified by world leaders to the present day.

Hawkins' "levels of consciousness", as previously mentioned, represent the degree of awareness the person has achieved, so that what seems like truth or reality varies with the level to which one's consciousness is aligned. As a culture we are brought up to be contentious to varying degrees, defending our egos and taking offence if we feel slighted.

Viewed from a higher perspective, Hawkins says "to be offended signifies that a one is defended which in itself signifies a clinging to untruth. Truth needs no defence and so is not defensive and has nothing to prove."[3] Hawkins says it is only when a person is open, non-defensive and willing to change that evolution to higher potential is possible.

**How We Can Evolve**

Our imperfection as a species suggests that we have a way to go on our evolutionary path. It is our hard-earned life experience and especially those reactions to life that are inspired with wisdom, which hold the possibility of actualizing the evolutionary potential within our psyches.

Apparently Ghandhi, when asked what he thought about Western civilization, replied that he thought it would be a good idea! Our so-called "civilized" lifestyle is basically lacking in civility.[4] It is difficult to apply the word "civilized" too much of our behaviour when the word itself is defined as an advanced state of human society whose people have reached a state of refinement and improvement in quality. One might well hope Teilhard de Chardin is correct when he asserts that there is a directional evolutionary force behind it all. If it were left to us to create our own reality with our own minds, our evolution as a species could be at risk. It is after all, our free will and the choices we ourselves have made, that have got us just where we are today, on the brink of disaster.

Bernard Haisch [and others] gives us some hope in spite of our sorry state. Haisch says that experience and observation have shown him that there is a self-creative, self-organizing characteristic of living things that indicates the influence of a higher order or Infinite Intelligence [we call God] which constrains our free will with basic laws.[5] Every experience advances us, he says, until we are transformed by that experience and return to the infinite Source from which we sprang. This from an astrophysicist sounds reassuringly similar to what the perennial philosophy has been telling us for centuries.

The force behind our evolution is showing us the need for a fundamental shift in consciousness in human kind. What is required is a period spent outside our comfort zone such as a Vision Quest, or regrettably perhaps a critical global crisis, to initiate our coming of age. When awareness does shift, we become transformed from helpless victim governed by the mind, ego and senses in a dangerous chaotic world, into a mature, intuitive, advanced being governed by wisdom and intuitive guidance and living in harmony with Divine forces.

It is our separation from the spiritual that underlies our present crisis with its impending ecological disaster that brings with it despair, disempowerment and powerlessness. A new vision is required that brings together the new science, spirituality and a sense of the sacred and that re-bonds us with the Earth. Matthew Fox and Rupert Sheldrake have made it clear that far from taking us back to the old animism and worship of nature gods, this spiritual vision of an evolutionary world where creativity is an on-going feature of the developing cosmos is a concept involving a "higher turn of the spiral" from primitive pantheism.

A new approach to both science and theology would mean connecting with life and creation in a different way and on a higher level. Nature is governed by what Sheldrake calls "morphic resonance", the influence of like-upon-like frequencies. This new approach would mean in practice that we choose to connect or resonate with creation in a different way, forming what they call an "I/thou relationship" with the earth and its inhabitants rather than the "I/It" relationship that characterises the outmoded mechanistic attitude.[6]

## Science Vs Mysticism

This solution requires that we drop our bias against mysticism as opposed to the great god, Science. We have lived for centuries in a patriarchal era which has put aside the feminine aspect [the mystical] and honoured only the fatherhood [rationality] and it is this imbalance that has impoverished our psyches. As a race we have tended to become soulless, cynical and despairing beings, and as such we are, ourselves, part of the cause of the present crisis.

Sheldrake and Fox remind us that during the 5,000 year period when mankind worshipped the mother goddess in Europe, anthropologists have found no evidence of military artifacts; instead only thousands of statuettes of pregnant women! In the great age of temple building, rather than construction of memorials to the fallen in great battles, the mother goddess Sophia ruled the universe with wisdom, compassion and justice for the poor and oppressed. In the twelfth century the cathedrals the powerful archetype of mother Mary was raised above all. Patriarchy replaced her with misogynous bishops.

The mysticism of which Sheldrake and Fox and many others speak, is at the heart of most of the religious traditions from the American First Nation peoples, to the Christian, Islamic, Jewish and Buddhist traditions. It is a "knowing" or sense of connection to a larger whole that ignites a mystical experience of Grace and "being graced". I felt it when surviving a potential plane crash in the Andes or the guidance that came when suffering the Dark Night of the Soul. It has graced the lives of so many of us and gifted us with a sense of connectedness with all that is.

A patriarchal attitude toward creation involves "power over" the environment and each other, while the feminine provides an I/thou-relation with all of life, one that involves honouring, reverencing and experiencing the sacredness within it. Meister Eckhart called this I-thou connectedness, "relating to the essence of everything that exists". This is the paradigm required; one where the division between the inner and outer, the subjective and objective, our minds and Creation is overcome. But in order to experience awe and reverence we must first quieten our minds and remove our conditioned sense of separation.

Creating a new vision has nothing to do with the ego's need to exploit the environment or create whatever we desire. Instead it means we expand our perception so that it is more inclusive, more responsible and more spiritually mature. The stumbling block in this process is the level of consciousness from which we are co-creating. As mentioned previously, many of us operate from an ego level that is governed by a limited, self-cherishing, power-based and defensive perspective, valuing one-upmanship and winning at all cost.

Brugh Joy also believes we must become unhooked from our limited conditioning before a new future becomes possible. He suggests we use what Buddhists call "beginner's mind", free of preconceived ideas beliefs and positionalities, to begin to understand the insanity that lies within our current world view. It is our conditioned beliefs and the idea that these beliefs represent reality that are the source of our wars, ecological destruction, genocide, incarceration, torture, crusades and jihads, all that is abominable about the human condition.

It is our conditioning, too, that prevents us realising the magnificence of our creative potential.

## 15. THE NEXT STEP ON OUR EVOLUTIONARY PATH

Another supercharged moment occurred when flying from Vancouver to Miami in Florida. The passenger area became suffused with fumes appearing to come from somewhere at the rear of the plane. As the fumes intensified, the plane turned and changed direction. We were told simply that we were making an "unscheduled" landing in Salt Lake City, but as we approached the airfield, the stewardesses gave us instructions in emergency landing procedures.

My husband, Ian and I were sitting by an exit over the wings so we were given special instructions for helping others exit the plane in an emergency. Afraid as I was, my recollection of the swift exit via escape chute from wing exit provided some sense of security. It was then we were instructed in the way to open the door above the wing, walk out to the end of the wing, and then how one of us was to help passengers to jump with their feet together to the tarmac while the other jumped down and aided those who jumped!

At this point the enormity of what was being asked was drowned by the fact that we about to touch the runway. If the smell indicated some electrical fault with the landing equipment as some suggested, the events that would occur in a few moment's time would one way or another convincingly prove that theory.

We made a blissfully uneventful landing accompanied by emergency vehicles of several kinds. It was as we walked briskly from the plane, however, that I chanced to look up at the wing. It towered above us at a height that indicated that a jump from that wing could in all probability have caused injury, perhaps severe injury. Where was the escape chute I had so eagerly anticipated?

As we toasted our survival with a double-shot cappuccino I felt again that exhilarating sense of relief. A peak moment of extreme fear and anticipation followed by exhilaration and heightened senses of sight, hearing and even smell, and certainly of wellbeing and gratitude.

The peak experiences I faced both in North and South America, while bringing an enormous high and a reassessment of my life as its gift, were both quickly forgotten as life progressed. While a peak experience may temporarily breach the crack in the hypnotic, conditioned egg of our existence, the task remains to take that peak awareness and make it part of ordinary experience. "The process of enlightenment . . ." Brugh Joy says, ". . . is the process of acquiring the ability to expand that tiny sphere of awareness to include more and more . . . a larger and larger portion of the large sphere." [1] Joy considers the expanded states of Beingness the natural ones, and the so-called "normal" state, the profoundly abnormal one.

Jack Kornfield says it is Buddhist philosophy that has the ability to take us to the next step in our evolutionary capacity to see beyond our limited perception. The self that clings to limited experience, he says, solidifies itself within that attractor field. This perpetuates suffering and counter-evolutionary mentalities. [2] The frequencies of judgment and hate draw to us events and people that mirror these frequencies.

Compassion, however, carries the charge that removes these negative attractor fields and allows the greater frequency potentials of subtle thought and emotion to arise. Buddhism provides a course for our lives, a guideline for self-awareness and living with compassion. Applying such principles does not mean necessarily that one should become a Buddhist but much can be gained from their "how to do it" approach which many religions do not so specifically provide.

The act of conscious intention leads us on the first step toward conscious evolution. Awareness itself is not static but fluid and expansive state. Brugh Joy gives the example of Edgar Cayce who could see people at a distance while in a trance state but while in ordinary consciousness he suffered all the normal constraints. From which state we choose to experience life is our individual choice, but it is incumbent on us to remember that our destiny is bound within that state.

In order to achieve an expanded state, it is essential to deeply experience the higher aspects of Love and compassion and to practice applying these in everyday living. We do this by using awareness to become conscious of the demons within our psyche that attempt to drag us in the opposite direction. Compassionate actions accompanied by generosity of spirit are the attractor fields whereby evolution occurs within the human psyche.

Joy believes that as the number of people attaining higher levels of awareness increases, a critical intensity will be achieved and collective consciousness will be shifted in a type of hundredth monkey effect, or in scientific speak, a phase transition will occur. It is this paradigm shift within the global human psyche that will be our salvation.

This is the really good news. Joy says it is this global paradigm shift rather than a worldwide catastrophe which is the essence of the cataclysmic climate prophecies that are rife in current thought. Current events do foreshadow a revolution of the most astonishing proportions but the upheavals reflect the conversion required in the mental, emotional and spiritual plane rather than social disorder. In his words, "I sense the approach of a psychological earthquake the magnitude of which has not been experienced in the human awareness for millennia and may not have been experienced in the human awareness ever before." [3]

Professor Tiller refers to the "higher qualities" being required within the human psyche as a feasible way to gradually raise the earth to a new level of potentiality and possibility. He believes this level could only be sustained and kept stable at that higher level by an elevated and loving human consciousness expressed as greater kindness, tolerance, understanding and empathy with our fellow human beings.

Because the brain becomes habituated into a way of thinking and behaving which cannot be easily dislodged, the overall requirement is that we must practice the new way by actually living it. The process of shifting our level of consciousness involves nothing less than replacing one way of thinking with another. All the nerve cells in the brain are interconnected so by consciously building sufficient numbers of elevated experiences in the neural receptors, negative thoughts and delusions become dismantled and new neural pathways consistently established.

Dominant and destructive behaviour can be re-wired to become more compassionate and life-enhancing responses. A long-term study of both human and primate, mainly baboon, health and wellbeing and their effect on brain chemistry, was carried out by Robert Sapolski, Neurobiologist and Primatologist at Stanford University.

He found that status in society had a huge effect on the mental and physical wellbeing of members of both human and baboon societies, the lowest ranking members, those most picked on and least rewarded for effort, had stress-related illness to the point of death. The dominant, highest ranking males in the baboon troop had the least stress even though seldom liked by troop members, while the lower members supported and groomed one another. Power-struggles were carried down the line of status and the lowest were tormented to such a degree, life became intolerable. Sapolski decided at this stage, after thirty years studying them, he did not actually like baboons.

The study then takes an extraordinary turn. His first-studied baboon troop was affected by tuberculosis after contact with human waste. It was the dominant males who stood alone and ate the major share who succumbed first and most were exterminated by the illness. Only the less dominant and more effeminate male baboons survived along with most of the females. His conclusion was that it was the mutual support and grooming among these baboons that improved their immune systems and allowed them to survive. The baboon society from that time changed. Feminine values became the dominant ones. Compassion, support and equality were the new rules and incoming males had to be taught these rules in order to remain. The wellbeing and overall health of the whole troop was improved by what at first appeared to be a disaster.[4]

Humans live in a stress-filled and in many ways a dog-eat-dog society where lower status is associated with increased stress and illness, both mental and physical, including damage to the brain cells. Will disaster be what it takes to bring us to a new way of living, just as it did with the baboons? Are we too attached to power over others, climbing the social ladder, and putting others down in order to improve our own self-image, to choose a higher level of consciousness over these goals? Are we capable of changing our attitudes and adopting a more harmonious lifestyle which will enhance our wellbeing?

Currently a war is raging in the Ukraine that, with its live coverage of atrocities, is horrifying television watchers throughout the world. In fact there appears to be little room for optimism at this time in the history of the planet. However, beneath the horror and tragedy, the inhumanity and the suffering, something else is revealing itself. It is the resilience and compassionate response of the victims, not occasionally but en masse and over long periods. Along with all the brutality and callousness that some members of the human race are capable of inflicting, there resides in others a more worthy, nobler side, despite the offenders' atrocities. It would be simpler to save oneself, to become caught up in hate and vengefulness. But instead they step into danger to help a stranger. Disasters, horrific as they are, uncover heroism and compassion that times of plenty keep hidden beneath comfort and complacency, and tragically in some are never expressed. At a time of crisis over much of the globe, my faith in human nature is restored by the nobility of spirit that arises, against all odds, in the hearts of mankind. My sense is now that our compassionate side will win in the end. Our evolution toward higher consciousness is possible.

During times of crisis, we make choices, some selfish, others heroic. In our daily lives we can bring attention to the quality of our choices which gradually builds new neural programs to increase the refinement of our mind. These new neural pathways have the capability of developing our capacity to acquire higher consciousness.

The benefits are many; greater possibilities and capabilities, more adaptability, a greater ability to feel compassion for another as well as the recognition that power over others is devolutionary. As a result we come to recognise and respect others as part of a larger connectedness with all that is. In other words this higher awareness gradually replaces the "I win, you lose" narcissistic mentality with an increased awareness and a greater ability to love as moment by moment we replace habitual behaviour with higher choices. This continual "bootstrap" process is the way we grow spiritually.

## We Are Capable of Attitude Change, But Will We?

A recognition of other people and all living things as part of our larger identity is possibly the most critical step in the spiritual process that can escalate mankind to another level. We could build a better world by directing our attention to the quality rather than the sheer quantity of our thoughts and actions. Behaviour that fosters mutual support and altruistic mind-sets enrich us as well as benefitting the larger-than-human community.

If we are to change in a meaningful way, an alternate model or paradigm needs to be provided to enable us to see beyond our current world view that power and control reign supreme. As we have seen, in order for the fully integrated world view to prosper, we must learn to develop a understanding of unconditional selflessness and to apply it in all areas of our lives **"The meaning of life has something to do with realizing that our essence is perfect love, then going on to live our lives upon that truth, experiencing each day as a miracle and every act as sacred.** "Human beings know far more than they allow themselves to know; there is a kind of knowledge of life which they reject, although it is born into them; it is built into them."[5]

The new paradigm involves redefining the purpose of existence to one of integration, development, harmony and order. A new paradigm is part of the evolutionary drive on this planet toward a more ordered, mature and perfected state, the one that is required to convert the world from the Tipping Point to the Turning Point.

Many scientists have come to this same conclusion. Albert Szent-Gyorgyi, as we know, defined negative entropy or syntropy as an innate drive in living matter to perfect itself. On the psychological level he spoke of a drive towards synthesis, towards growth, wholeness and self-perfection. Griffith concluded that syntropy makes life's integrative, cooperative, loving and selfless meaning inescapable. Arthur Koestler, Hungarian industrialist, inventor and author also referred to the active striving of living matter towards its optimal evolutionary potential.

Mysticism in all its forms has pointed to this same universal goal throughout the centuries. Sir James Darling, renowned Australian educator concludes, "The scientist can no more deny or devaluate the truths of spiritual experience than the theologian can neglect the truths of science; and the two truths must be reconcilable" [6]

Wayne Dyer believes the world is even now experiencing a "phase transition". As in quantum physics, when enough electrons line up within an atom to form a position, all the rest automatically line up in a similar fashion, so too humanity is beginning to bring about a phase transition as sufficient number of human beings begin to form a critical mass perhaps prompted by the threat of global warming. "The miracle of cooperation rather than competition is the beginning phase of a transition to a more positive, safe, loving world." [7]

Our lives are interwoven with all life but mankind is uniquely in the position to husband and nurture God's creation. Our individual mental, emotional and spiritual mind-sets have a critical impact on the survival of all living things. Our personal responsibility is great and we need to be up to the task. Our planet is crying. Its creatures are crying, the human race is crying. The question becomes; have we found sufficient compassion in our hearts to overcome the menace created by our own minds?

Melchizedek puts it well when he says, "Ultimately, we can only find a harmonious way through this passage by finding our hearts. That is the message. Why? Because our heart is naturally tuned to the unity of all of life and therefore lives in respect of it." By finding genuine compassion, we access genuine wisdom.

Meditation is a means by which a radical shift of orientation occurs. At the meditative level, we have access to the subtle energetic blueprints of transformation. The goal of meditation, according to sages of all time, is the realization of the inner Self, one's own unique unity with the creative process. It is the Insight Intelligence, accessible through a meditative state, that allows us to develop the higher levels of nonphysical thought we must have for complete development and wellbeing.

### Apocalypse Now

Daniel Pinchbeck believes Revelation's "apocalypse" in the Bible represents the drastic series of trials when humanity, after transforming itself through those trials, becomes fit to enter into direct relationship with Divinity. Surely this is precisely the task of contemporary humanity, confronted as it is with a possible apocalypse of its own making. "Right now", he says, "we are being forced to witness the shadow of the psyche projected into material form through systemic misuse of technology." ..... "If the shadows appear to be growing darker it is because the light that casts them is getting brighter."

We might need to face the fact that we live during a period in our history that is forcing us to evolve at high speed. Breakdown and breakthrough may happen simultaneously. In Pinchbeck's words, a higher consciousness cannot take place in a "subconscious murk." [8] It is through a sincere Vision Quest into this murk and the systematic de-conditioning from negative programming and ego-centric goals that we begin to separate ourselves from the darker side of our character.

Just as a projector creates a version of a film's own ephemeral reality on a screen, our shadow side is projected outwards onto everything and everyone in its vicinity tainting the whole environment with its own delusionary beliefs and values. The transformation of consciousness therefore requires not only personal work but actual application of the new consciousness to our *whole* environment; the ecological, political, psychological, commercial, technological and spiritual aspects of reality. Transformation involves an initiation, an inauguration into a deeper dimension of our being sufficient to bring harmony to the practicalities of our lives a troubled world.

We are in the midst of a world in transition, involving all beings in an evolving planet. In a world where both Heaven and Hell have been equally the human experience throughout the ages and where Science now speaks of parallel universes allowing both to be experienced simultaneously, a cataclysmic period for some of mankind could equally bring a period for others of the attainment of our highest potential.

The way to open the new field of potential within us, is to incorporate spirituality into the very substance of our everyday lives thus allowing the sacred to become a living reality through our renewed perception thus allowing the sacred to become a living reality through our renewed perception. [9]

A rise in people's consciousness level is the means by which the universe becomes more aware and more compassionate. We can choose a more expansive reality perhaps in a similar way as the shaman, Don Yuan who taught Carlos Castaneda to see a "second reality" through choosing his right brain rather than the left.

New beliefs and practices throughout the ages each created a more ethical and principled outcome. Examples of these include large movements such as changes from feudal or autocratic regimes to greater social equality, the bringing down of the Berlin wall, and the abolition of apartheid. On a local scale it includes non-polluting cars, smoke-free restaurants, work places, buses and planes, new solar electric homes, recycling and movements concerned with the environment. The application of the new paradigm applies to all social and environmental organisations.

From environmentalist Thomas Berry's perspective, "The capacity for ordered self-development, for self-expression, must be considered as a pervasive psychic dimension of the universe from the beginning. All particular life-systems must integrate their being and their functioning within this larger complex of mutually dependent Earth systems." [10]

"Earth spirituality" is the word used by Claudia von Werlhof who believes we need a different culture; the word "culture" used in its original meaning of "nurturing". "Right now, we nurture machines rather than community. We nurture violence rather than love." Movements such as Earth spirituality would provide a far-reaching global change in mankind's way of thinking, feeling and acting so that we convert patriarchal separation, superiority and control to a more encompassing connectedness relating the material, the mind and the soul. [11]

Ultimately, ecological awareness is spiritual or religious awareness. Bill Reed expresses the same idea when he quotes Fritjof Capra, "When the concept of the human spirit is understood as the mode of consciousness in which the individual feels a sense of belonging, of connectedness, to the cosmos as a whole, it becomes clear that ecological awareness is spiritual in its deepest essence... Somehow we need to become whole again – connected and integrated." [12]

The innate drive in living matter to perfect itself, coupled with human intent, takes us toward order and wholeness and brings integrative, cooperative, nurturing, meaning and purpose to all areas of existence and evolution to humanity itself.

**We can become the saviours our planet has been waiting for.**

## 16. OUR RESPONSIBILITY TO THE PLANET

"Watch your breath!" came the instruction yet again, thrusting me back from my daydream into reality. I had been sitting cross-legged with 40-50 others at an ashram outside Sydney for several long days, beginning impossibly early in the morning and finishing about 9pm.

Just why I would put myself through such a punishing regimen with its restrictions and work schedules during my precious school holidays was something I was to ask myself repeatedly as I imagined the more obvious alternative; sun-filled beaches and parties with friends. What on earth was I going to tell my fellow teachers when I returned for first term and they related their overseas trips and wonderful family camping trips? Why would an ostensibly sane person spend holiday time sitting in silence from 4am until after dark watching her breath? I suspected it be fairly difficult for me to maintain any air of credibility with my colleagues after this.

Sadly, watching my breath did not bring me enlightenment. To be honest, it appeared to do nothing at all. Nor did watching my thoughts; after all my thoughts, to me at least, were fairly engaging so that before long I became hooked into the content and forgot the watching-bit altogether. The chanting which involved repeating lines in Sanskrit twisted my tongue and scrambled my mind, but no expansive awareness followed. I was obviously a complete failure at this meditation thing.

Then followed a practice called Yoga Nidra in which most of the group took the opportunity for a welcome nap just after lunch. My system was far too "hyped" to fall asleep in public and so I actually did as I was told. I was asked to experience my big toe, a request which might have appeared bizarre to anyone who had not just spent two weeks watching my breath. Having suspended disbelief on a grand scale already, I decided to give it my best shot and actually felt my toe enlivened by my attention!

The practice took my awareness from the outside sounds into every minute part of my body so that I felt each part from the inside. Gradually I could take that attention to my whole leg, and eventually my whole body. At the point when the orange-clad yoga instructor said, "whole body awareness", energy began to fill my entire system. I was acutely aware of a flow of beingness that spread throughout my body and extended outside of it like the light from a torch penetrating a certain distance into the darkness.

I could watch this energy as an observer but also simultaneously as a participant. The awareness filled my mind, my body and my whole energy field. Over time I began to be aware of the feeling of various mental and emotional states within different parts of my body. Years later, when chronic fatigue overtook my overwrought system and nervous grinding energy became my way of functioning in the absence of my natural life-force, I could calm it only with this new body awareness. When overwhelm threatens my equilibrium, I quieten my energy field. When crisis hits, I face it with mindfulness. Being all too human this is a process still in progress.

As mentioned previously, mindfulness brings a higher frequency bandwidth to consciousness so that the sacredness of life experience is able to be perceived more intimately.

Something very extraordinary occurs when one uses mindfulness and expanded awareness. Many of us resemble the James Joyce character, Mr. Duffy, who lived "a short distance from his body".

A lot of humanity lives an even longer distance from an awareness of their innermost motivations and agendas. Most of us escape instead into the realms of the impersonal and relatively unchallenging mental realm. This escape route serves beautifully to avoid any real soul-searching and camouflages awareness and its insights, replacing them with intellectualism, thus leaving the ego free to play its tricks unnoticed and unchecked.

Moments of silence are in short supply in our modern life. We surround ourselves with sound every moment of the day. We crave distraction and avoid aloneness at all cost. We carry devices for keeping in contact at all times. We demand to be entertained and our senses tempted with the latest gismo or fashion. We are seldom quiet enough to allow subtle levels to reach us.

We strive to achieve, we struggle to get ahead, or we just struggle out of habit. We live on adrenalin. We are addicted to drama, both in our own lives and vicariously in sitcoms and movies which include large doses of violence, infidelity and macho states. Incidentally, Buddhists say mindfulness allows meditators to be less aggressive and confrontational. There is also more harmony in the homes of meditators.[1]

We began this book with a disheartening look at the state of the world but found instead that the current situation could become a catalyst for change ushering in the need for a quest into the cause of our lamentable behaviour as a species and the possible cures for it.

The cause, we found, centres largely on society itself and the level of consciousness it engenders. Society is shaped by culture and the media into negative and self-indulgent attractor fields of power, wealth and celebrity, all cosmeticized so as to hold us within that particular perceptual level. The level of consciousness of the population becomes entrained at a minimal level where violence and entitlement become accepted as the norm. Our goals become ephemeral, our attitudes centred on self-indulgence and our sense of purpose too easily diminishes to mimicking those with status and celebrity. The result of all this is that we lost a sense of the sacred in our lives.

The cure involves a change of consciousness that has the potential to take us from Tipping Point to Turning Point. The question then becomes; how do we take a Vision Quest into our own minds? The answer lies in awareness and scrutiny of our innermost thoughts. Many researchers believe that much of humanity behaves as if hatred, anger and vengefulness were an inevitable part of our life experience. It is a sad assessment of humanity.

However we can choose to inhabit a level of consciousness where contention and discord with others are no longer our prevailing patterns. With awareness we can choose not to succumb, not to react, and to simply allow others to live out their melodramas without being personally touched, except by compassionate acceptance. Other peoples' opinions, issues and reactions are their own, and in fact none of our business.

**The trick is to not trust your mind. The mind is a wily and slippery customer.** It is best not to take its word to be the absolute truth. Snares lurk within the psyche; jealousy fear, anxiety, anger and unworthiness are persistent and very stubborn. These emotions are saboteurs. Refuse them the time of day. Our shadow side contains powerful energies, so just thinking nice positive thoughts will not budge them. It takes large doses of self-discipline coupled with courage, self-honesty, humility and a large measure of determination to shift those demons.

The stories we use to justify our actions are also devious and too easily convince us they represent who we actually are as people. They subtly transform themselves into our identity. We become "the sufferer", "the know-all", "the wise being", "the manipulator", "the goody-goody", the one who was hard-done-by in childhood, the one with the high status, the misunderstood or unheard one, the one who gains respect because of power, position or wealth. Socially these stories may have been useful; spiritually they keep us embedded in ego defences and self-deception.

We transcend the lower levels only when we have developed the ability to refuse to identify with them, but instead, to watch the mind's chicanery.

Peace is reached through undergoing the turbulence; wisdom is found through seeing through our game-playing ego states; radiant presence is achieved through transcending the inner shadow, and self-mastery is found through overcoming our limiting mind-games.

**Every word we utter, every action we take, every comment we make about another reveals our own quality as a person. Even the words we fail to utter; the kindness we neglect to show, the failure to give just a little more to another, reveal who we are as a person. Generosity of spirit is a virtue not many of us can claim as our own.**

It is the innermost essence of a person, his or her level of consciousness, that produces a more beautiful human being, or a lovely atmosphere to be part of, a person who attracts harmony, loving relations, peace and joy, is the kind of person who can lift the spirits of all those around them. It has been said that the high vibrations of an enlightened person can balance a large number of those at low levels of consciousness.

With so many of the world's population below the level of integrity, according to David Hawkins, the planet owes a lot to those few who rise to a higher level. Ultimately it is the quality of our "Being" that will save the world.

Several mind-awareness practices can aid the person in seeing through their mind's stories and its trickery; Vipassana, Yoga, Qi Gong, heart-centred awareness practices, silence and stillness in Nature. Certain psychological practices such as Gestalt and Rational Emotive therapy can also assist with attaining psychological health.

The ideal would be to integrate psychology, brain neurology, perennial philosophy, sociology, ecology, and interrelationship skills, together with ever-expanding scientific knowledge, diverse, inconsistent and even antagonistic as they may appear to be. This would expand our perspective, our perception and our world view from the diverse to a more united approach.

Evidence from scientific studies, for instance, suggests that contemplative and meditative practice, usually associated with ancient philosophies, actually change the brain. Self-induced states of tranquility are associated with more positive emotions and even improved immune response. Tibetan monks use mindfulness to control their body processes and even to raise their body temperature sufficiently to melt snow in their Himalayan monastery grounds. Incidentally, Lincoln Hall, who at age 50 survived 36 hours at 28,000 feet in the Death Zone of Everest, attributes his survival to Buddhist mindfulness and breathing practice.

Einstein's view was that the mind creates ways of separating things and concentrating on differences so we miss the oneness. One of his greatest trademarks was a complete commitment to seeing the universe as one seamless whole. [2]

How different the world would be if we related to it from this perspective.

Martin Buber, an Austrian-born Jewish philosopher and honorary professor at the University of Frankfurt, best known for his philosophy of dialogue, a form of religious existentialism, claims that in the main we relate to other people as 'I-It' as objects utterly separate from –and hence considered subordinate to - ourselves. When relating as 'I and Thou' we lose the sense of self and other as separate and enter a perception of oneness. Before Christ, the ancient Essenes believed that compassion transcends our doing, and that it is through compassion that you allow yourself to become. This is just as valid today, and just as little adhered to. The virtue, decency and kindness you most wish to have in your life, you must first become in your life.

**The moral virtues we need to acquire are the old-fashioned ones of prudence or moral intelligence, humility, sagacity, insight-informed conscience, rigorous honesty, using past experience to change our ways, ability to find harmony and balance. The bottom line is we need to learn to live love.**

Love is so integral to this transformation that we must cut our addiction to holding grievances, spite, violence, control and winning at all cost. Henry Longfellow wrote "If we could read the secret history of our enemies we should find in each man's life sorrow and suffering enough to disarm all hostility."

Within the duality of existence, order and chaos exist side by side. Transcendence must be embodied in dualistic physicality and we ourselves are those embodiments. Our purpose is to take the opposites and in a moment by moment interchange, convert duality into unity; we need to let go but be conscious that there are times when it is appropriate to take a stand. We must love unconditionally but be sure to establish personal boundaries. We should give with a generosity of spirit while not allowing ourselves to be over-ridden and downtrodden by those who seek to control and manipulate. Transformation is clearly not for wimps.

Basically our progress requires what appears to be a contradictory mix of personal attainments; determination and surrender, mental awareness and intuition, focussed intention and trust, strength of will and a sense of humour, insight/ inspiration and grounding in the material world.

What we have formerly seen as our strengths; power, position, status, wealth, intellect, can in fact become our undoing. It is the "soft" side which we hide, our hearts and our ability to love and to intuit, that turn out to be the secret of our salvation. According to Epigenetics, the emotions we choose determine the actual patterning of our DNA through the vibrational frequency of our level of consciousness. Our beliefs and attitudes, the manner in which we resolve our emotional challenges, whether by compassion or confrontation, affirm life or deny it.

It is infinite, conscious intelligence of the zero point field that is the origin of matter and the laws of nature, and because we are part of this universal field of consciousness, we shape our world by the love or malice, the compassion or indifference we bring to it. Elevated and loving human consciousness is the key requirement. We create our collective future via our thoughts, attitudes and actions. It is the love, compassion, devotion and intent that can elicit the 'unseen' assistance of the universe. In this troubled world we could do with a large dose of that assistance.

Assistance, as we have learned, comes from the universal field in the form of Insight intelligence which allows us to break free of our conditioning, and is the key to perceiving the world from a higher perspective of interconnectedness. It enables us to become a species that has the wisdom to caretake the earth rather than simply the mental capacity to dominate it. This field of universal intelligence or zero point field, with its characteristics of being all loving, all-knowing, all powerful creator energy may be equated with the Source, Intelligent Design or in other words, the Divine!

Syntropy with this universal field causes organisms to continue to "try to perfect themselves", providing a blueprint for our evolution as a species, enabling humanity to reach much higher levels of development. If the evolutionary drive on this planet is toward a more ordered, mature and perfected state then as part of this evolutionary drive, our participation in this process is vital.

The new evolutionary paradigm involves redefining the purpose of our existence to one of integration, development, harmony and order. With the knowledge that we are part of the evolution and survival of all life, it becomes obvious that we are being called by the emergency of our present global situation to take the personal responsibility these critical circumstances are demanding of us. We must, as a matter of urgency, apply the higher energies of cooperation rather than the accepted energies of competition in order to transform a critical situation into a more positive, safe, loving world.

The journey we have taken to find a solution for our world on the brink of catastrophe brings us to an incredible conclusion; we must acknowledge that each one of us, along with all of creation, is a participant in God's evolutionary plan. We are incredibly, God's expression in human form of His/Her/Its very being. Our obligation the higher energy of Creation carries with it is that we ground the archetype of interconnectedness and compassion into the physical world. This raises our collective level of consciousness and inspires us to live with love and wisdom, dedicated to reverence for life in all its forms.

Far from destroying God, science for the first time is proving the existence of an all-encompassing God-force, [or in scientific terms, the Zero Point Field] by demonstrating that a higher Consciousness is the bottom line in all of creation. There need no longer be two truths i.e. science and religion.

The meaning of our existence and our highest purpose as individuals is therefore to take our rightful place in the evolution of the human species and, with our unique position among all species, assume responsibility for the wellbeing of the planet. Becoming aware of the sacredness of life brings with it a personal commitment to the collective consciousness and to the part we play in an interconnected global village. As Marshall McLuhan said, "There are no passengers on spaceship Earth. We are all crew." [3]

By deepening our identification with all life-forms, with ecosystems and with the planet herself we begin to discover within the "ecological self" – the broader and deeper self that is a natural member in the more-than-human community, its environment, its climate and its wildlife as well as all races of humans. If we were then to extend these higher vibrational frequencies into our diverse academic institutions, into the business world, into politics, into inter-relationships within the community and the country as a whole, how different our world would be.

Our rite of passage in the underworld has brought us to a Vision that can initiate us into our coming of age, our true adulthood as a species. We have travelled to the depths of our being on this Vision Quest for our rightful place in our troubled world and returned with the astounding awareness that each of us has the God-given power and the awesome responsibility to contribute to the evolution of mankind, all its creatures and its habitat. The truth we carry back from our Vision Quest is that by opening to an inner knowing of interconnectedness, and empowered by connection with the Divine, we have the capacity to restore harmony and peace in a world on the brink of catastrophe.

It is easy to think that as individuals we cannot make a difference to something as overwhelming as the evolution of humanity and the survival of the planet. In fact, it is impossible for any one of us not to make a difference.

The kind of difference we make, however, depends on the quality of our own individual level of consciousness. [4]

# OTHER BOOKS BY THIS AUTHOR

We trust you have gained real value from reading:
*"Healing the Earth - What's Love Got To Do With It?"*
It is Book 2 in the set: *Tipping Point or Turning Point?*

The accompanying Book 1 is:
*"Compassion and Other Earth Changing Options -*
*Where Spirituality and Frontier Science Meet"*
Both bring fresh insightful perspectives, are complimentary in their content, and can be read singly, or in either order.

You might also like to read Sandra's first book
*"Where is God When Times are Tough? – The Journey to Inner Wisdom – Finding Answers to Life Challenges"*
This inspirational book will assist you in transcending troubles and connecting with your own inner wisdom and insight.

Check out the details at Amazon or their associated outlets.
All three books are available in printed paperback and Kindle.

Cheers
Maleny Press
December 2020

Healing the Earth – What's Love Got to Do With It?

# REFERENCES

### Introduction

1. Wheatley, Margaret, "Leadership and the New Science: Discovering Order in a Chaotic World"

### Chapter 1 - Perception and Violence

1. Boteach, Rabbi Shmuley, "The Private Adam – Becoming a Hero in a Selfish Age" Hodder & Stoughton, London, 2003, p22.
2. Hawkins, David R, "Power Vs Force, - An Anatomy of Consciousness. The Hidden Determinants of Human Behaviour", Hay House, CA, 1987, p237.
3. Zukav, Gary, "The Seat of the Soul. - An Inspiring Vision of Humanity's Spiritual Destiny"
4. Hawkins, David R, "Power Vs Force - An Anatomy of Consciousness. The Hidden Determinants of Human Behaviour" Hay House, CA, 1987 p181
5. Jung, Carl, "Civilization in Transition - The Collected Works of Carl Jung", Vol. 10,
6. Kornfield, Jack, "The Wise Heart - Buddhist Psychology for the West", Rider, 2008, p208

### Chapter 2 - Evolution Gone Wrong?

1. Hodgkinson, Sandra "Where is God When Times are Tough?", Spiritual Soulcraft, 2011
2. Haisch, Bernard "The God Theory, Universes, Zero-Point Fields, and What's Behind it All", Red Wheel/Weiser, p55
3. Bailey, Paul "Think of an Elephant - Combining Science and Spirituality for a Better Life", Watkins Publishing, London, 2007 p 328-330, p342-44

4. Hodgkinson, Sandra May, "Compassion and Other Earth Changing Options" Maleny Press, 2020, p77-p121
5. Pearce, Joseph Chilton, "The Magical Child" Bantam Books, 1986, p52.
6. ibid p20.

## Chapter 3 – The Third Matrix

1. Bailey, Paul, "Think of an Elephant - Combining Science and Spirituality for a Better Life", Watkins Publishing, London, 2007, p133.
2. Pearce, Joseph Chilton, "The Magical Child" Bantam Books, 1986 P260
3. ibid p226.
4 Talbot, Michael "The Holographic Universe", Harper Collins, London, 1991, p133.

## Chapter 4 - The Hitch in Evolution's Plan

1. Beauregard, Mario and O'Leary, Denyse, "The Spiritual Brain - A Neuroscientist's Case for the Existence of the Soul" Harper Collins, 2007, pp289-295.
2. Pearce, Joseph Chilton, "Evolution's End", Harper Collins, New York, 1992, p152.
3. Bailey, Paul, "Think of an Elephant - Combining Science and Spirituality for a Better Life", Watkins Pub, London, 2007, p229-30, 57.
4. Kornfield, Jack, "The Wise Heart - Buddhist Psychology for the West", Rider, 2008 p312.
5. Redfield, James and Murphy, Michael "God and the Evolving Universe", Jeremy P Tarcher/Putnam, NY, 2001, p151.
6. ibid p152.
7. Pearce, Joseph Chilton, "Evolution's End", Harper Collins, New York, 1992, p213.
8. Pearce, Joseph Chilton, "The Bond of Power", Elsevier-Dutton, NY, 1981, p59.
9. Haisch, Bernard, PhD, "The God Theory - Zero-point Fields, Universes, and What's Behind it All", Red Wheel/Weiser, S.F. 2006, p46.
10. McTaggard, Lynne "The Field - The Quest for the Secret Force of the Universe", Harper, New York, 2008, p226.

## Chapter 5 – Patriarchy

1. Pearce Joseph Chilton, "Magical Child" Bantam Books, 1986, p264.

2. Griffith, Jeremy, "A Species in Denial" FHA Pub, South Australia, 2003, p109.
3. Jung, Carl "Civilization in Transition", The Collected Works of Carl Jung, Vol. 10.
4. Griffith, Jeremy, "A Species in Denial" FHA Pub, South Australia, 2003, p236.
5. ibid p339.
6. Hope, Barbara "Patriarchy - A State of War", Class of Nonviolence. Lesson Five. Essay Five.
7. Pinchbeck, Daniel '2012- The Return of Quetzalcoatl" Penguin, NY 2007, p. 369.
8. Hope, Barbara, "Patriarchy: A State of War", Class of Nonviolence.

## Chapter 6 - Force Vs Empathy

1. Darwin, Charles, "Descent of Man" Gibson Square Books, London, 2003, p619.
2. 'The Thoughts of the Emperor m Aurelius Antoninus" 1869.

## Chapter 7 - Mankind's Right to Dominate?

1. Galdakis, Birute "Reflections of Eden - My Life with the Orangutans of Borneo", Victor Gollancz, London, 1995, p253.
2. ibid p245.
3. ibid p356.
4. ibid p253.
5. Fouts, Roger "Next of Kin - What Chimpanzees Have Taught Me About Who We Are", p154
6. Darwin, Charles, "The Descent of Man", Gibson Square Books, London, 2003, p126.

## Chapter 8 - Kinship With All Life

1. Weber, Bill & Vedder, Amy, "In the Kingdom of Gorillas - The Quest to Save Rwanda's Mountain Gorillas." Aurum Press, UK, 2001, p361.
2. Hawkins, David R "Power Vs Force - An Anatomy of Consciousness. The Hidden Determinants of Human Behaviour." Hay House, CA, 1987, p233.
3. Darwin, Charles, "The Descent of Man" Gibson Square Books, London, 2003, p123.

4. ibid p101.
5. ibid p103.
6. Lovelock, James "The Revenge of Gaia - Why the Earth is Fighting Back and How We Can Still Save Humanity", Penguin, London 2006, Intro by Sir Crispin Tickell.
7. Plotkin, Bill "Soulcraft - Crossing into the Mysteries of Nature and the Psyche" New World Library, California, 2003, p118.
8. Macy, Joanna, "Joanna Macy on the Great Turning: A Shift in Consciousness" - Summer 2006: 5,000 Years of Empire, p42.
9. Moore, Thomas, "Re-Enchantment of Everyday Life", Harper Collins, 1996, p29.

**Chapter 9 - Sacredness Within Nature**

1. Sanchez, Victor, "The Teachings of Don Carlos", Bear & Company Publishing, Santa Fe, 1995, p148.
2. McTaggard, Lynne, "The Field - The Quest for the Secret Force of the Universe", Harper, New York, 2008, p6.
3. Lovelock, James, "The Revenge of Gaia - Why the Earth is Fighting Back and How We Can Still Save Humanity" Penguin, London, 2006, p154.
4. Darwin, Charles, "Descent of Man", Gibson Square Books, London, 2003, p94.
5. Davies, Paul, "The Mind of God - The Scientific Basis for a Rational World", p128.
6. Braden, Gregg, "The Isaiah Effect - Decoding the Lost Science of Prayer and Prophecy", p109.
7. Haught, John, "God After Darwin - A Theology of Evolution", Westview Press, 2007.
8. Haisch, Bernard, "The God Theory - Universes, Zero-Point Fields, and What's Behind It All" Red Wheel/Weiser, SF, 2006, p36-7.
9. Tiller, William, "Conscious Acts of creation: the emergence of a new physics"p14
10. Haisch, Bernard,, "The God Theory - Universes, Zero-Point Fields, and What's Behind it All", Red Wheel/Weiser, San Francisco, 2006, p95.
11. Chopra, Deepak, "How to Know God - The Soul's Journey into the Mystery of Mysteries" p71.
12. Haisch, Bernard, "The God Theory - Universes, Zero-Point Fields, and What's Behind it All" Red Wheel/Weiser, San Francisco,, 2006, p12.
13. Moore, Thomas, "Re-Enchantment of Everyday Life", Harper Collins, 1996, p4, 29

14. Lovelock, James, "The Revenge of Gaia - Why the Earth is Fighting Back and How We Can Still Save Humanity", Penguin, London, 2006, p138.
15. Zukav, Gary, 'The Seat of the Soul - An Inspiring Vision of Humanity's Spiritual Destiny"
16. Dyer, Wayne W, "Real Magic - Creating Miracles in Everyday Life.".
17. Sanchez, Victor, "The Teachings of Don Carlos", Bear & Company Publishing Santa Fe, 1995, p23.
18. Sardello, Robert, "Love and the World - A Guide to Conscious Soul Practice" Lindisfarne Books, MA, 2001, p148.
19. Moore, Thomas, "Re-Enchantment of Everyday Life", Harper Collins, 1996 p343.
20. Sardello, Robert, "Love and the World- A Guide to Conscious Soul Practice", Lindisfarne Books, MA, 2001, p41.
21. Quoted in "What the Bleep - A Study Guide", Institute of Noetic Sciences, Captured Light Industries.
22. "Ecosystems – Man's Interference", "Tiger Conservation is People Conservation", LifeForce Charitable Trust.
23. Silva, Freddie, "The Secrets in the Fields - The Science and Mysticism of Crop Circles" Hampton Roads Publishing Co, Charlottesville, VA, 2002, p169.

### Chapter 10 - "Creating Your Own Reality"

1. Moss, Richard, "The Second Miracle - Intimacy, Spirituality and Conscious Relationships" Celestial Arts, Berkeley, CA, p101.
2. Chopra, Deepak, "How to Know God - The Soul's Journey into the Mystery of Mysteries", p131.
3. Hawkins, David R, "The Eye of the I - From Which Nothing is Hidden", Veritas Sedona, Arizona, 2001, p141.
4. Sheldrake, Rupert and Fox, Matthew "Natural Grace – Dialogues on Science and Spirituality," Bloomsbury Publishing, London, 1996, p101.
5. Tiller, William A., Prof. Emeritus, Stanford Univ., "Conscious Acts of Creation: the Emergence of a New Physics"
6. "What the Bleep. Study Guide", Institute of Noetic Sciences, Captured Light Industries.
7. Hawkins, David R, "The Hidden Determinants of Human Behaviour.", Hay House, CA, 1987 p79-80

8. Jourdain, Stephen, "Radical Awakening - Cutting Through the Conditioned Mind". Dialogues with Stephen Jourdain, Inner Directions Publishing, California 2001 p40,79.
9. Hawkins, David R MD., PhD, "Reality and Subjectivity", Veritas Publishing, West Sedona, USA, 2003, p222.
10. "What the Bleep - Study Guide", Institute of Noetic Sciences, Captured Light Industries.
11. Ornish, Dr. Dean, "Love and Survival –The Scientific Basis for the Healing Power of Intimacy", Random House, Australia, 1999.
12. Joy, Brugh, MD "Joy's Way - An Introduction to the Potentials for Healing with Body Energies", JP Tarcher, Los Angeles, 1979.

**Chapter 11 - The Perspective of the Perceiver**

1. Bailey, Paul, "Think of an Elephant - Combining Science and Spirituality for a Better Life", Watkins Publishing, London, 2007, p74
2. Lawton, Ian, "Genesis Unveiled, The Lost Wisdom of our forgotten Ancestors", Virgin Books, London, 2003, p300.
3. Braden, Gregg, "The Isaiah Effect - Decoding the Lost Science of Prayer and Prophecy" Crown Publications, 2002.

**Chapter 12 - Love and the Heart**

1. Pearce, Joseph Chilton "Expressing Life's Wisdom - Nurturing Heart-Brain Development Starting With Infants", Interview by Chris Mercogliano and Kim Debus, Journal of Family Life, Volume 5, No.1, 1999.
2. Fraser, Peter H & Massey, Harry "Decoding the Human Body-field - The New Science of Information as Medicine", Inner Traditions/Healing Arts Press, 2008, p6.
3. McTaggard, Lynne, "The Field - The Quest for the Secret Force of the Universe" Harper, New York, 2008, p36.
4. Pearsall, Paul, "The Heart's Code - Tapping the Wisdom and Power of Our Heart Energy" Bantam, Australia, 1999, p2-3.
5. ibid p192 quoting from Margharita Laski's 'Ecstasy: a Study of Some Secular and Religious Experiences".
6. "What the Bleep - Study Guide", Institute of Noetic Sciences, Captured Light Industries.
7. Melchizedek, Drunvalo, "Living From the Heart", Light Technology Publishing, Flagstaff, US, 2003, p112.

8. Pearsall, Paul, "The Heart's Code, - Tapping the Wisdom and Power of Our Heart Energy", Bantam, Australia, 1999, p71.
9. Ornish, Dr. Dean, "Love and Survival –The Scientific Basis for the Healing Power of Intimacy", Random House, Australia, 1999, p. 190.

## Chapter 13 - The Critical Role of Parenting

1. Fraser, Peter & Massey, Harry, "Decoding the Human Body Field: The New Science of Information as Medicine". Inner Traditions/Healing Arts Press, 2008, p4.
2. Ornish, Dr. Dean, "Love and Survival –The Scientific Basis for the Healing Power of Intimacy", Random House, Australia, 1999, p223.
3. Bradshaw, John, "Reclaiming Virtue", Bantam Books, New York, May 2009 p94
4. Lipton, Bruce, "Sabotaging Ourselves with Limiting Subconscious Programs", Living Now Magazine, August /September 2009, p4.
5. Carter, Christine, "Survival of the Kindest - Social Scientists Build Case for 'Survival of the Kindest", Science Daily, Dec. 9, 2009.
6 Griffith, Jeremy, "A Species in Denial", FHA Publishing, South Australia, 2003.

## Chapter 14 - What the World Needs Now

1. Kornfield, Jack, "A Path with Heart - A Guide through the Perils and Promises of Spiritual Life". Bantam, NY, 1994, p295.
2. Lovelock, James, "The Revenge of Gaia - Why the Earth is Fighting Back – and How We Can Still Save Humanity.", Penguin, London, 2006, p13.
3. Hawkins, David R, "I - Reality and Subjectivity" Veritas Publishing, West Sedona, USA, 2003, p312.
4. The Prince of Wales "The Modern Curse that Divides Us from Nature" - Times Online, The Times, November 27, 2008.
5. Haisch, Bernard PhD, "The God Theory, Universes, Zero-Point Fields, and What's Behind it All", Red Wheel/Weiser, San Francisco, 2006, p. 43-46.
6. Sheldrake, Rupert, & Fox, Matthew, "Conscious Acts of Creation - Natural Grace – Dialogues on Science and Spirituality", Bloomsbury Pub, London, 1996, p23.

## Chapter 15 - Next Step on our Evolutionary Path

1. Joy, Brugh, MD "Joy's Way - An Introduction to the Potentials for Healing with Body Energies", JP Tarcher, Los Angeles, 1979, p113.
2. Kornfield, Jack, "The Wise Heart - Buddhist Psychology for the West", Rider, 2008, p65-66.
3. Joy, Brugh, MD "Joy's Way - An Introduction to the Potentials for Healing with Body Energies", JP Tarcher, Los Angeles, 1979, p279.
4. National Geographic documentary, Sept. 2008.
5. Griffith, Jeremy, "A Species in Denial" FHA Publishing, South Australia, 2003, p262.
6. ibid p413.
7. Dyer, Dr. Wayne W, "Real Magic - Creating Miracles in Everyday Life"
8. Pinchbeck, Daniel, "2012- The Return of Quetzalcoatl", Penguin, NY 2007 p109.
9. Pearce, Joseph Chilton, "The Biology of Transcendence - A Blueprint of the Human Spirit," Park Street Press, Vermont, 2002, p201.
10. Berry, Thomas, "Twelve Principles for Understanding the Universe and the Role of the Human in the Universe Process", p3.
11. Von Werlhof, Claudia, "The Interconnectedness of All Being - A New Spirituality for a New Civilization".
12. Reed, Bill, "Evolving Towards Unity", Sustainability and Interconnectedness, February 2005.

## Chapter 16 - Our Responsibility to the Planet

1. McTaggard, Lynne "The Field -The Quest for the Secret force of the Universe", Harper, New York, 2008, p212.
2. Ornish, Dr. Dean, "Love and Survival - The Scientific Basis for the Healing Power of Intimacy", Random House, Australia, 1999, p123, 177
3. Bailey, Paul, "Think of An Elephant - Combining Science and Spirituality for a Better Life", Watkins Publishing, London, 2007, p377.
4. ibid, p381.

www.ingramcontent.com/pod-product-compliance
Lightning Source LLC
Chambersburg PA
CBHW020649220526
45464CB00001B/350